CL[illegible] c. at the University [illegible] work in industry he [illegible] and Fairwater Sch[illegible] d of the Upper Sch[illegible] sive School, and then [illegible] Comprehensive Sch[illegible] Headmaster of Caerleon Comprehensive School, he was for many years an evening lecturer at Gwent College of Higher Education. Chief Examiner in General Science and Chemistry for two C.S.E. Boards, he also examines at G.C.E. Ordinary Level.

key facts

GCE O-Level Passbooks

BIOLOGY, R. Whitaker, B.Sc. and J. M. Kelly, B.Sc., M.I.Biol.

ECONOMICS, J. E. Waszek, B.Sc.(Econ.).

ENGLISH LANGUAGE, Robert L. Wilson, M.A.

FRENCH, G. Butler, B.A.

GEOGRAPHY, R. Knowles, M.A.

HISTORY (Political and Constitutional, 1815-1951), L. James, B.A., M.Litt.

HISTORY (Social and Economic, 1815-1951), M. C. James, B.A.

MODERN MATHEMATICS, A. J. Sly, B.A.

PHYSICS, B. P. Brindle, B.Sc.

REGIONAL GEOGRAPHY (British Isles), R. Bryant, B.A. and R. Knowles, M.A.

TECHNICAL DRAWING, P. Barnett, D.S.C., M.C.C. Ed., Adv. Dip. Ed.

GCE O-Level Passbook

Chemistry

C. W. Lapham, M.Sc.

key facts

Published by Charles Letts Books Ltd
London, Edinburgh and New York

Previously published 1976 by Intercontinental Book Productions

Published 1983 by Charles Letts Books Ltd
Diary House, Borough Road, London SE1 1DW

2nd edition, 1st impression 9.82.15

Printed by Charles Letts (Scotland) Ltd
ISBN 0 85097 539 5

Contents

Introduction

This book is intended for those students taking the examination leading to the General Certificate of Education at Ordinary Level, and the Certificate of Secondary Education. In addition, certain parts will be useful to students following Certificate of Extended Education (CEE) courses.

The contents include the majority of the topics listed in the various syllabuses of the regional examining boards. S.I. (International System) units are used throughout, and in particular the new nomenclature relating to atomic weights, molecular weights and molar masses is included. This new nomenclature will be used by some boards for the first time in the 1976 examination papers.

Subject areas such as equations, electrolysis and the mole concept with which students often experience difficulty are dealt with in detail.

Students will find when consulting careers advisers and literature that chemistry qualifications, both in their own right and allied with other subjects, are of great importance in many walks of life. Students of home economics, for example, are realising that a knowledge of chemistry to at least O-level standard is essential to their studies, and an increasing number of colleges of domestic science are insisting upon chemistry qualifications. A study of A-level biology requires a knowledge of O-level chemistry if students are to benefit in full from the course. For those students with a general interest in the subject this book should provide a good background knowledge upon which to build.

Acknowledgements

To my wife Brenda, Karen, Lee and Jeffrey for their encouragement and tolerance; to Mrs Susan Thomas for her hard work in typing and checking the manuscript, and Mrs Pat Bosanko for her splendid work in the revision of the book; to colleagues past and present for their advice and help.

Chapter 1
The States of Matter

All materials are described by scientists as **matter**, and can be categorised as being **solids, liquids or gases**, these being known as the **three states of matter**. The planet earth consists of these three states in the form of **land, sea and air**.

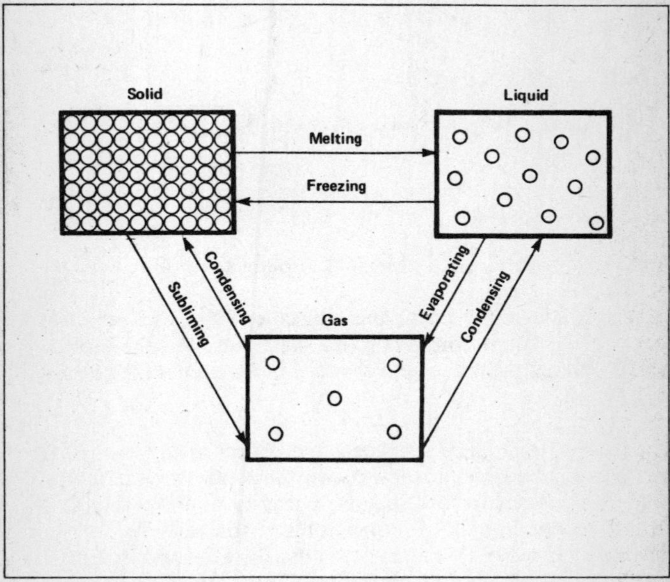

Figure 1. The three states of matter and their interconversion

When ice is **heated** it **melts** to form the **liquid** water which in turn with sufficient heat energy will **boil** to produce the **gas** called steam. The reverse processes involve the withdrawal of heat energy. Melting, freezing, condensing and evaporating are well-known processes. However, some solids when heated do not undergo the usual solid-to-liquid-to-gas stages but change on heating straight into a gas, omitting the liquid stage. Such a process is called **subliming**, and is well illustrated by the action of heat upon the compound ammonium chloride or sal ammoniac.

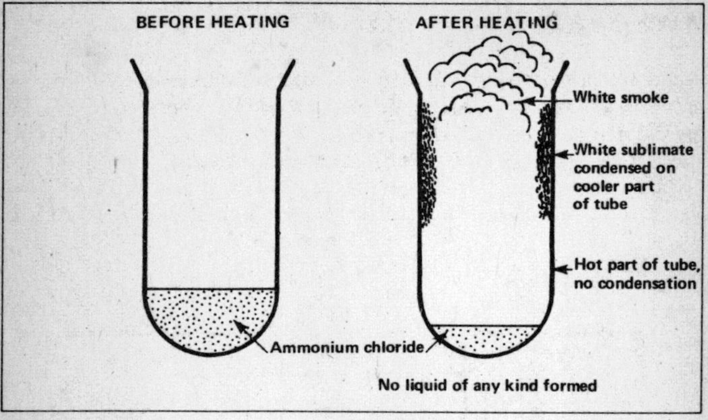

Figure 2. Sublimation

Another common substance which sublimes is iodine.

In the solid state the atoms and molecules are very closely packed together. In liquids there is more space between the molecules, while the greatest space between molecules occurs in the gaseous state.

The major difference between the three categories is their respective molecular movements. In the solid state the molecules take up positions to form regular patterns held together by very strong forces, e.g. crystal structures. Molecules in solids can move very slowly but movement is usually confined to vibrations within the fixed pattern or lattice. When a solid is heated the molecules acquire energy from the source of heat resulting in stronger vibrations until eventually the molecules break free from the forces holding them in position. When this occurs the **melting point** of the solid has been reached or exceeded.

Liquids are less dense than solids. The molecules of a liquid, being further apart, can therefore move around much more freely. However, forces of attraction do hold the molecules of a liquid together. At the surface of the liquid these forces of attraction are much weaker than in the body of the liquid, making it easier for molecules at the surface to leave the liquid, When molecules escape in this way the process of **evaporation** is said to be taking place.

Heating a liquid, or giving the molecules in a liquid greater energy, **increases the rate of evaporation** because the heat energy makes the movement of the molecules much faster and stronger, enabling more and more to leave at the surface of the liquid. This is evident when water is heated. When the boiling point of a liquid is reached the molecules break away completely from the attractive forces in the liquid and a gas is formed. Tabulated below are the general properties of solids, liquids and gases.

Property	Solid	Liquid	Gas
Volume and shape	Definite volumes and shapes	Definite volume – takes the shape of the container	Volume not definite – takes the shape of the container
Particle motion	Very slow vibrations	Moderate movement	Fast random movement in all directions
Density	High	Medium	Low

The theory that molecules are always in some kind of constant movement, whether simply vibrating around a fixed position (as in solids) or moving fast at random in all directions, is known as the **kinetic theory of matter** ('kine' meaning movement). On the basis of the kinetic theory various phenomena can be explained, such as those described in the remainder of this chapter.

Pressure of a gas

Gas pressure arises from the bombardment by gas molecules of the walls of the vessel in which it is contained. If the temperature of a given volume of a gas is increased the pressure is also increased.

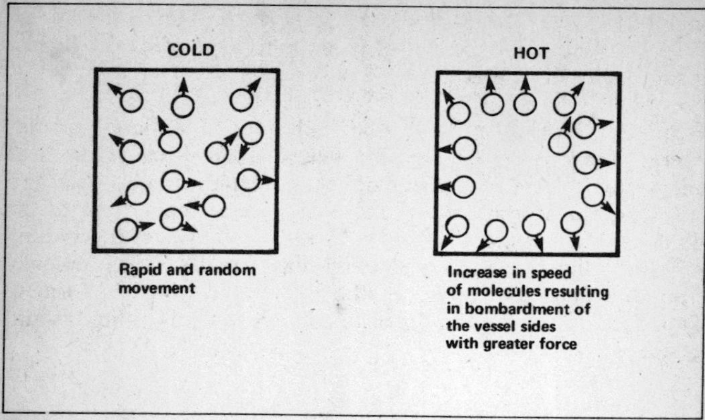

Figure 3. Increase of pressure of a gas with increase of temperature

If a gas is heated then the molecules move faster and farther apart – that is, the volume of the gas increases. This important point leads to **Charles' law** which states that **the volume of a given mass of gas is directly proportional to its absolute temperature as long as the pressure remains constant.** E.g. If the temperature of a given mass of gas is doubled then its volume is doubled.

i.e. $V \propto T$ V = volume
$V = k.T.$ T = absolute temperature
$\frac{V}{T} = k$ k = constant

For two such situations then

$$\frac{V_1}{T_1} = \frac{V_2}{T_2}$$

Note In order to convert centigrade or Celsius into absolute (Kelvin) simply add to the centigrade temperature 273°.
E.g. 15°C would be 288K
$15 + 273 = 288.$

Another important law relating to gases is **Boyle's law** which states that **the volume of a fixed mass of gas is inversely**

proportional to the pressure provided the temperature remains constant.
E.g. If the plunger in a closed gas syringe is pushed inwards, the gas in the syringe is compressed, or its volume is decreased as the pressure is increased.

$$V \propto \frac{1}{P} \qquad V = \text{volume}$$

$$\text{i.e. } V = \frac{k}{P} \qquad P = \text{pressure}$$

$$\text{or } PV = k \qquad k = \text{constant}$$

For two such situations then

$$P_1V_1 = P_2V_2$$

Charles' law and Boyle's law can be combined to give the Gas Equation

$$\frac{P_1V_1}{T_1} = \frac{P_2V_2}{T_2}$$

Expansion

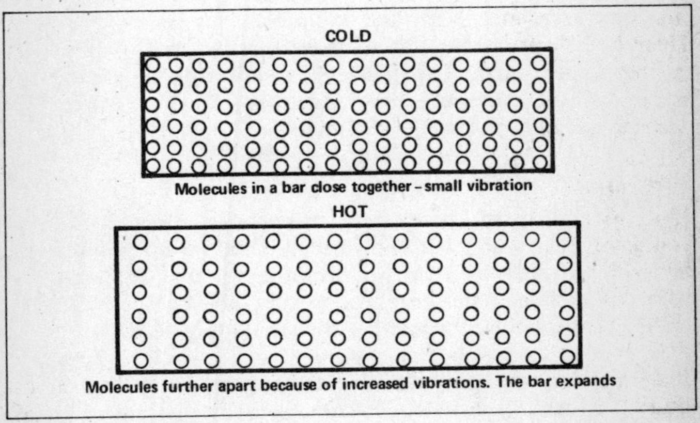

Figure 4. Expansion in a solid

The diagrams above should be self-explanatory. When the bar is cold the molecular movement is slow and the molecules are close together. Upon heating the molecular movement increases, the molecules move farther apart and the bar expands.

Conduction

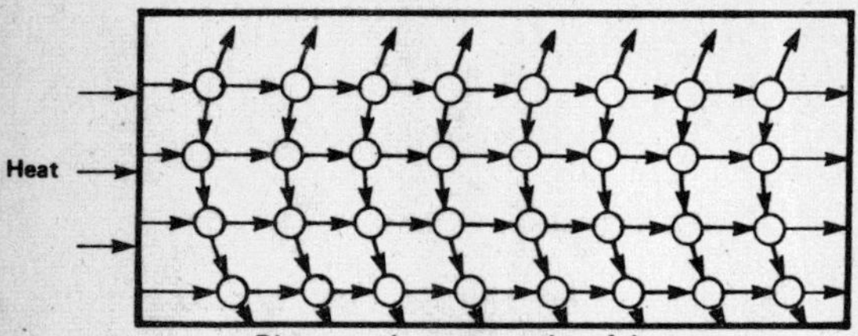

Diagrammatic representation of the process of conduction. The molecules acquire heat energy, increase speed and strike molecules next to them, transferring heat energy along the substance

Figure 5. Conduction

If one end of a piece of metal is heated, the molecules and free electrons associated with the molecules increase their speed. These free electrons are able to move quickly passing on energy to the cooler parts. The greater the heating the greater the energy of the molecules and electrons, and the faster the transference of heat through the metal.

Diffusion

If a small drop of the element bromine is placed in a large container, sealed and left undisturbed, within a few minutes the jar will be filled with brown bromine vapour. Bromine is a volatile liquid and the evaporation that takes place will result in the gas spreading throughout the containing vessel.

If a small crystal of potassium permanganate is placed at the bottom of a large beaker of water, colour striations will start moving through the liquid until eventually the whole liquid is uniformly pink.

Both the above experiments illustrate the fact that the molecules of liquids and gases move freely. This process is called **diffusion**.

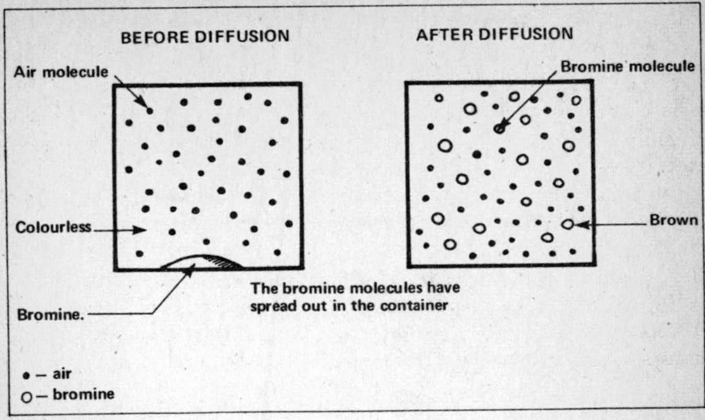

Figure 6. Diffusion

Diffusion in liquids is slower than in gases. There are many everyday examples of diffusion ranging from cooking odours emanating from a kitchen into other rooms of a building to the pleasant smell of perfume from soaps and cosmetics. Many many more examples of diffusion occur naturally.

Diffusion of a gas or liquid is the spreading of the substance owing to the spontaneous movement of its molecules from one place to another.

Some gases can diffuse more quickly than others, hydrogen (the lightest of all gases) being the fastest. Graham's law of diffusion states that **the rate of diffusion** of a gas is inversely proportional to the square root of its density.

$$R \propto \frac{1}{\sqrt{D}} \qquad R = \text{rate}$$

$$R = k.\frac{1}{\sqrt{D}} \qquad \begin{array}{l} D = \text{density} \\ k = \text{constant} \end{array}$$

If the rates of diffusion of two gases are R_1 and R_2 respectively and their densities are D_1 and D_2 then

$$\frac{R_1}{R_2} = \sqrt{\frac{D_2}{D_1}}$$

Remember that D_1 and D_2 are **densities**, not vapour densities. The density of a gas is proportional to its vapour density which in turn is proportional to its molecular weight, thus

$$\frac{R_1}{R_2} = \sqrt{\frac{M_2}{M_1}}$$

where M_1 and M_2 are the molecular weights of the gases. Hence the molecular weight of a gas can be determined by comparing its rate of diffusion with that of a gas whose molecular weight is known, i.e. if any three of the four variables in the above equation are known it is quite easy to calculate the fourth.

E.g. 400 cm^3 of a gas took 40 seconds to diffuse through a porous pot. The same volume of hydrogen took 10 seconds to diffuse under the same conditions. From this information calculate the molecular weight of the gas.

According to Graham's law $\frac{R_1}{R_2} = \sqrt{\frac{M_2}{M_1}}$

the rate of diffusion of each gas per second is

$$\text{(i)} \quad \frac{400}{10} = 40 \text{ cm}^3 \text{ per second}$$

$$\text{(ii)} \quad \frac{400}{40} = 10 \text{ cm}^3 \text{ per second}$$

the molecular weight of the gas hydrogen (H_2) = 2

$$\therefore R_1 = 40$$
$$R_2 = 10$$
$$M_1 = 2$$

Substituting into the expression

$$\frac{R_1}{R_2} = \sqrt{\frac{M_2}{M_1}}$$

$$\frac{40}{10} = \sqrt{\frac{M_2}{2}}$$

$$\frac{M_2}{2} = \frac{40^2}{10^2}$$

$$M_2 = 32$$

Standard temperature and pressure (s.t.p.)
Standard temperature is 273 K.
Standard pressure is 760 mm of mercury.

To convert to s.t.p.
This is best demonstrated by an actual calculation.
E.g. The volume of a gas at 17°C and 740 mm of mercury is 250 cm^3. What will be the volume of the gas at s.t.p.?

Temperature considerations

(i) 17°C is 290 K

(ii) The volume of the gas at 273 K will be **less than** that at 290 K because volume is proportional to absolute temperature.

$$\text{(Charles' law) volume} = 250 \times \frac{273}{290}$$

Pressure considerations
At 760 mm of mercury the volume of the gas would be less than it would be at 740 mm because volume is inversely proportional to pressure (Boyle's law).

$$\text{volume} = 250 \times \frac{740}{760}$$

Combining the two:

$$\text{volume at s.t.p.} = 250 \times \frac{273}{290} \times \frac{740}{760}$$

which can be evaluated using logarithms.
Note Remember that 760 and 273 are never both in the denominator or numerator. One is in the numerator and the other in the denominator. Time can be saved by remembering this fact, because it is only necessary to consider either temperature **or** pressure and then write the full expression accordingly. E.g. in the above calculation it was deduced that the 273 was in the numerator, hence the 760 must be in the denominator and vice versa.

Brownian movement
Robert Brown, in 1827, noticed the continuous irregular movement of pollen grains suspended in water. This irregular movement occurs because the water molecules, moving with kinetic energy, bombard the pollen from all sides producing movement. A similar situation occurs when highly illuminated tobacco

smoke is observed to be moving in an irregular haphazard motion because of the bombardment of the smoke molecules by the molecules of air.

Phenomena such as diffusion, conduction, expansion and pressure all contribute evidence to the idea of matter being made up of very tiny particles. This theory is known as the **particulate theory of matter**.

Key terms

States of matter Solid, liquid and gas exemplified by land, sea and air.
Sublimation The process that occurs when a heated solid goes straight to the gaseous state without first passing through the liquid state, e.g. ammonium chloride, iodine, solid CO_2.
Kinetic theory The theory of molecules being in some kind of constant movement.
Charles' law The volume of a given mass of gas is directly proportional to its absolute temperature as long as the pressure remains constant, i.e. $V \propto T$.
Boyle's law The volume of a fixed mass of gas is inversely proportional to the pressure (at constant temperature), i.e. $V \propto \frac{1}{P}$.
Diffusion The spreading of a substance (a gas or liquid) owing to the movement of its molecules from one place to another.
Standard temperature and pressure (s.t.p.) The standard to which a given volume of gas is referred is 0°C, i.e. 273 K and 760 mm of mercury.

Chapter 2
Mixture and Solutions

A **mixture** consists of two or more substances, either compounds or elements, which are not combined together chemically: e.g. air is a mixture of the gases oxygen, nitrogen, carbon dioxide and inert gases, etc. Oxygen, nitrogen and the inert gases are elements, whereas carbon dioxide is made up of the elements carbon and oxygen, chemically combined together to form a **compound**.

Differences between compounds and mixtures

	Mixtures	Compounds
(i)	A mixture can have any composition	A compound has a fixed composition
(ii)	The components can easily be separated from each other by physical or mechanical means	The components can only be separated by chemical means
(iii)	The properties are those of each component or intermediate between those of the component parts	The properties are usually quite different from those of the component parts
(iv)	In making a mixture there is usually no heat change involved	There is often a large heat change in the formation of compounds

Mixtures can be categorised into four main groups:
(i) mixtures of solids, e.g. salt and sand.
(ii) mixtures of solids and liquids, e.g. sea water.
(iii) mixtures of liquids, e.g. crude oil.
(iv) mixtures of gases, e.g. air.

Separation of mixtures

Differences in the properties of the various constituents of each mixture can serve a useful purpose, e.g. to separate a mixture of iron filings and sulphur use is made of the facts that (i) iron is magnetic and (ii) sulphur is soluble in carbon disulphide. In other instances differences in such properties as solubility, density, boiling points etc. could facilitate separation.

(i) **Mixture of solids**

Solution can be a very useful method for separation of mixtures of solids. It is first necessary to find a suitable **solvent** that will dissolve one or other of the solids in the mixture; the most common and widely used solvent is water, but others used in certain instances include acetone, tetrachloromethane (carbon tetrachloride), xylene, and ethanol.

The mixture is shaken up with the solvent until one of the constituents is in **solution.** The constituent which has dissolved is called the **solute**. Therefore

$$\text{solute} + \text{solvent} = \text{solution}$$

The **insoluble** component is then separated from the solution either by **filtering** in the usual filtration apparatus or by centrifuging and decanting the **supernatant liquid**. The insoluble solid is washed, dried and subsequently stored.

Note Filtration will take place very much faster when the solution is **hot**.

The **soluble** component can be recovered from the solution (filtrate) by crystallisation or evaporating to dryness by heating.

Note The latter method, evaporating to dryness, can only be used when the solute is not affected by heat.

(ii) **Mixture of solid and liquid**

The second part of (i) applies here too. If the solid is insoluble separation is achieved by filtration or centrifuging and decanting. If the solid is dissolved in the liquid and a solution has been formed, the solid can be recovered by evaporation of the liquid, e.g. copper (II) sulphate can be obtained from a solution of copper (II) sulphate by **slow evaporation** or crystallisation. **Fast evaporation** or evaporating to dryness is the other method available provided the solute is unaffected by heat, e.g. salt can be obtained from a salt solution simply by heating the solution until all the water has boiled away as steam leaving behind salt. Lead nitrate could not be obtained from an aqueous solution by evaporating to dryness because it breaks down on heating.

(iii) **Mixture of liquids**

Differences in the respective boiling points of the liquids are used to achieve separation in this case. The liquids are separated by **fractional distillation**. The liquid with the lowest boiling point distils first and is condensed, followed by the second liquid and so on. A fractionating column is used in conjunction with the usual distillation apparatus.

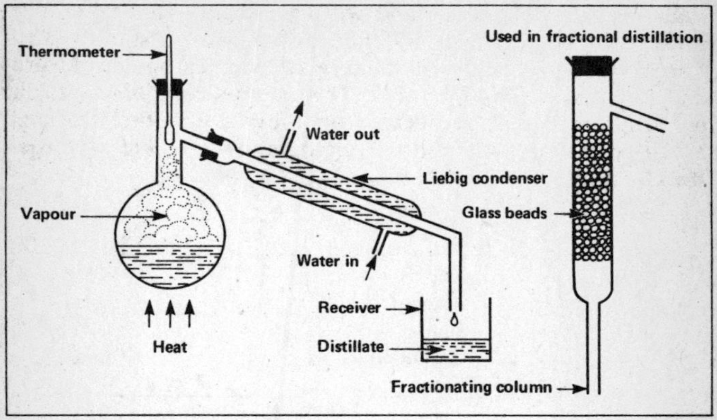

Figure 7. Distillation apparatus

Crude oil is a mixture of liquids and is separated into the different constituents by **fractional distillation** on a huge scale. Below is a list of the products obtained from crude oil, or petroleum as it is often known.

Fraction	**Derivatives**	**Distillation temp. °C**
Gasoline	Petrol, naphtha	up to 200°C
Kerosene	Paraffin oil, jet fuel	up to 300°C
Light oils	Diesel and gas oils	300 – 330°C
Lubricating oils	Liquid paraffin, paraffin wax	330 – 390°C
Bitumen	Heavy oils, asphalt	Residue

Liquids which do not mix completely with each other, e.g. oil and water, can be separated by pouring them into a separating funnel, allowing the two layers to separate out (most dense at the bottom) and then running off the lower layer into a container.

Chromatography

Two or more soluble solids (particularly if their solution is coloured) can be separated by chromatography. This process depends upon the fact that some solutions under the prevailing experimental conditions are adsorbed more easily than others. E.g. If the pigments from green leaves are dissolved in ether, and poured into a narrow glass column containing finely divided calcium carbonate, the top of the column will turn yellow while nearer the base a green band will be evident. This experiment shows that the pigment is made up of two substances, their separation being achieved because calcium carbonate adsorbs one more easily than the other. The separate components of the pigment can be easily recovered from the column. The experiment is an illustration of column chromatography or column **adsorption analysis.**

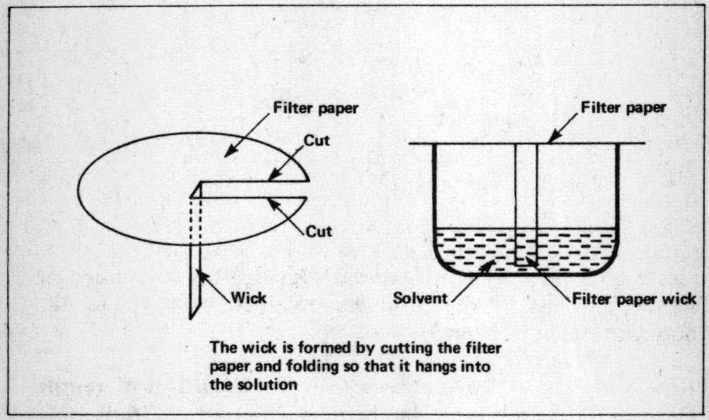

Figure 8. Paper chromatography

Paper chromatography

Here the liquid is drawn up the wick (figure 8) to the centre of the absorbent paper. From this it spreads out from the centre, the different substances in the liquid being absorbed at different rates resulting in a **chromatograph** consisting of a series of concentric rings.

A rough chromatograph can be obtained by placing a small drop of black ink in the centre of a piece of filter paper and allowing the ink to dry. Water is then run on to the centre of the paper. It spreads and carries the various dyes in the ink with it. A series of concentric coloured rings are obtained showing the composition of the ink.

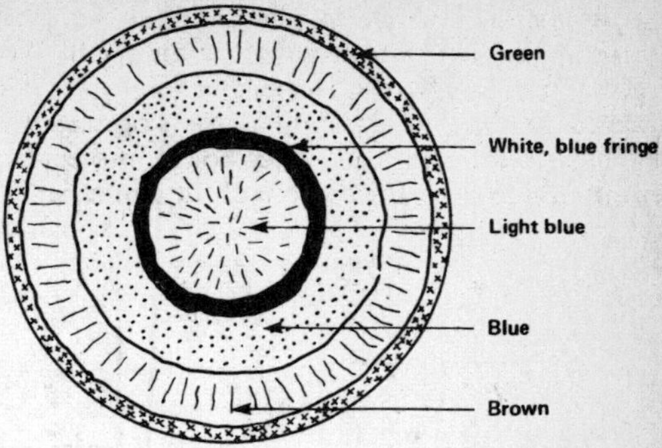

Figure 9. Chromatogram of black ink

(iv) **Mixture of gases**

Certain chemicals are capable of **absorbing** various gases, and use is made of this fact to separate gaseous mixtures, e.g. a mixture of carbon dioxide and oxygen could be separated by bubbling the mixture through a solution of caustic potash which would remove the carbon dioxide.

Air is the main source of oxygen and nitrogen for industrial use. The gases are separated from each other by fractional distillation of liquid air. Air is cooled and subjected to huge pressures which convert it into a liquid; the liquid air is fractionally distilled, nitrogen boiling off at $-196°C$. and oxygen at $-183°C$. The inert gases can be obtained from the remaining fraction.

Solutions

When common salt is added to water it eventually disappears, and is said to have dissolved in the water, or to be **soluble** in water. A solution of salt in water is thus formed. Salt, being the substance that dissolves, is called the **solute**, while water, the liquid into which the salt is dissolving, is called the **solvent**.

If powdered chalk is shaken up with water a cloudy liquid is produced called a **suspension**. If the liquid is allowed to stand for a short time the small white particles of chalk gravitate to the bottom of the container. The chalk does not dissolve but when shaken with water it is suspended in the water for a while. Some medicines must be shaken before use in order to disperse the fine insoluble particles into the liquid media.

The difference between a solution and a suspension is in the particle size. When a substance dissolves the extremely small particles cannot be detected, even with the most powerful microscope, while in a suspension the solid, although in a very fine state of subdivision, still consists of particles which are much too large to mix with the liquid molecules, and therefore eventually settle at the bottom of the container.

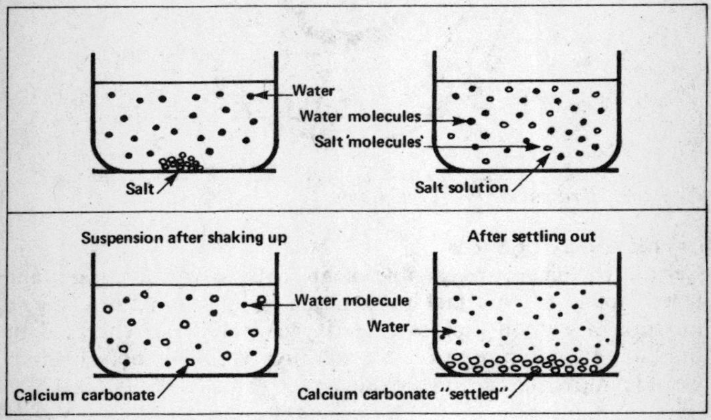

Figure 10. Solutions and suspensions

Metal **alloys** are **solid solutions,** e.g. British 10p and 50p coins, are made from an alloy of copper and nickel, the copper dissolving the nickel to form a solid solution of nickel in copper. Similarly, brass is a solid solution of zinc in copper. If an alloy which was a solid solution were to be examined under a very high-powered microscope, its individual constituents would not be resolvable.

Saturated solutions If more and more of a solute is added to a solvent, e.g. more and more salt is shaken up with water, a stage will be reached when no more salt will dissolve and excess solid will be visible at the bottom of the container. When this has occurred the solution is said to be **saturated**.

Note When a saturated solution is made it is important that the temperature remains constant.

Some substances are more soluble than others and their saturation points will vary. Temperature will also have a great effect on solubility. It is therefore necessary to define what is meant by solubility.

The **solubility** of a substance in a solvent at a given temperature is the number of grams of the substance that will dissolve in 100 grams of the solvent in the presence of undissolved solid to form a saturated solution.

Determination of solubility This can be done by making a saturated solution at the temperature (T°C) at which the solubility is to be found. A known volume, say 25 cm^3, of the solution is withdrawn, placed in a dry evaporating dish of known weight and the solution carefully evaporated to dryness. After cooling in a desiccator the dish and solid are weighed. The weight of solid can be calculated, hence the solubility.
E.g. 5g of solid were found to be dissolved in 25 cm^3 of water in the above experiment.

25 cm^3 of water dissolves 5g of solid

1 cm^3 of water would dissolve $\frac{5}{25}$ g of solid

100 cm^3 of water would dissolve $\frac{5}{25} \times 100$g of solid

$= 20$g

1 cm^3 of water weighs 1g

$\therefore$ 100g of water dissolves 20g of solid at T°C.

Solubility $=$ 20g at temperature T°C.

Solubility curves The solubility of most substances increases with temperature. If the solubility of a substance is determined at various temperatures then it is possible to plot the results on a solubility curve. Note the difference in the steepness of the curves in figure 11. Potassium nitrate, for instance, is far more soluble in hot water than potassium chloride, while sodium chloride shows hardly any variation. Using this information it is possible to separate solutions of salts from each other by making use of their different solubilities. This process is known as **fractional crystallisation**.

Recrystallisation Solid products may contain impurities when first made in the laboratory. They can be purified by being recrystallised from suitable solvents, e.g. the organic compound tri-iodomethane (iodoform) can be purified by dissolution in warm alcohol and subsequent recrystallisation. Crystals of pure iodoform are formed, the impurities remaining in the **mother liquor**.

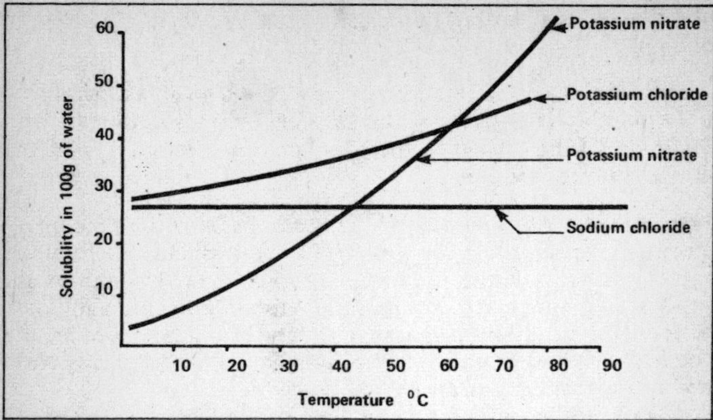

Figure 11. Solubility curves

Key terms

Mixture Two or more substances which are not chemically combined together, e.g. iron filings and sulphur.

Compound Two or more substances chemically combined together, e.g. sodium chloride.

Element A substance which cannot be split into anything simpler by chemical means, e.g. copper, silver, carbon, sulphur.

Solute The substance that dissolves when a solution is made, e.g. salt (solute) in a salt solution.

Solvent The substance that causes dissolving to take place when a solution is made, e.g. salt (solute) dissolves in water (solvent).

Solution The result of a solute dissolving in a solvent, e.g. solute (salt)+solvent (water)=solution.

Distillation The process of heating a liquid to above its boiling point in order to convert it to the gaseous state.

Chromatography The analysis of a mixture by selective absorption techniques.

Suspension A mixture of a liquid and a finely divided insoluble solid, 'held' by the liquid.

Alloy Two or more metals associated as a solid solution, compound or heterogeneous mixture.

Saturated solution Solution in which no more solute will dissolve at constant temperature.

Mother liquor The liquid associated with the crystal after crystallisation has taken place.

Chapter 3
Formulae and Equations

Atomic symbols

Atoms of elements are each represented by a symbol, e.g. iron by Fe, copper by Cu, etc. The symbols in each case represent **one atom** of the substance for which they stand – hence Fe denotes one atom of iron, Cu one atom of copper, and so on. Symbols for some of the other elements are given below.

Element	Symbol	Element	Symbol	Element	Symbol
Aluminium	Al	Hydrogen	H	Phosphorus	P
Argon	Ar	Iron	Fe	Potassium	K
Bromine	Br	Lead	Pb	Silver	Ag
Calcium	Ca	Magnesium	Mg	Sodium	Na
Carbon	C	Mercury	Hg	Sulphur	S
Chlorine	Cl	Nitrogen	N	Tin	Sn
Copper	Cu	Oxygen	O	Zinc	Zn

Where possible the first letter of the English name is used, however, where several elements begin with the same letter, e.g. **c**arbon, **c**opper, **c**hlorine, **c**alcium, C is used only once, for carbon, and another letter is added from the English or Latin name for the others, e.g. Ca – **ca**lcium, Cu – copper (from the Latin **cu**prum). Where two letters symbolise an element, the first should be a capital and the second a small letter, e.g. an atom of sodium would be represented by Na and **not** NA.

Formulae

A hundred and fifty years ago John Dalton would have said that when two atoms combine together a 'compound atom' is formed. Nowadays the so-called 'compound atoms' are usually referred to as molecules. E.g. H_2O represents one molecule of the compound water, which consists of two atoms of hydrogen, and one atom of oxygen.

All gaseous elements with the exception of the inert gases exist as pairs of atoms, e.g. O_2 represents one molecule of oxygen containing two atoms of oxygen combined together (note that 2.O represents two atoms of oxygen which are **not** combined together). Other gases which behave like oxygen are:

H_2 – hydrogen N_2 – nitrogen

Cl_2 – chlorine Br_2 – bromine (fuming liquid)

They are called di-atomic gases.

Radicals

Sulphuric acid has the formula H_2SO_4 and could be referred to as hydrogen sulphate. This acid gives rise to a series of compounds called **sulphates**, all containing the group **$-SO_4$**. The $-SO_4$ group cannot exist on its own and is called the **sulphate radical.** Similarly $-NO_3$ is the nitrate radical, $-PO_4$ the phosphate radical, while all ammonium compounds contain the NH_4- radical.

A **radical** could therefore be defined as a group of atoms which can exist in compounds but do not normally exist in their own right.

Valency or combining capacity

Atoms of elements combine in simple whole numbers, and the **valency** is a measure of the power of these atoms to combine with each other. The basis of this measure of **combining capacity** is **hydrogen** whose valency is taken as 1. Valency of an element is therefore the number of hydrogen atoms that one atom of the element will combine with or displace. The same reasoning can be applied to radicals.
E.g. the formula of sulphuric acid is H_2SO_4. This means that the sulphate radical $-SO_4$ is combined with two atoms of hydrogen, each having a valency of 1; the valency of $-SO_4$ must therefore be 2.

Some elements can have two valencies, e.g. iron can have a valency of 2 or 3, while copper can have a valency of 1 or 2. As a result there are two series of iron and copper compounds, iron (II) formerly called ferrous, iron (III) (ferric), copper (I) (cuprous), copper (II) (cupric). Examples of compounds are given below:

iron (II) sulphate	(ferrous sulphate)	$FeSO_4$
iron (III) sulphate	(ferric sulphate)	$Fe_2(SO_4)_3$
copper (I) oxide	(cuprous oxide)	Cu_2O
copper (II) oxide	(cupric oxide)	CuO

It is important to understand that the Roman numerals in brackets after the iron and copper refer to the **valency** of the respective atoms and **not** to the number of atoms in the compound. Some of the more common elements and radicals are:

Element	Symbol	Valency
Aluminium	Al	3
Bromine	Br	1
Calcium	Ca	2
Carbon	C	4
Chlorine	Cl	1
Copper	Cu	1 or 2
Hydrogen	H	1
Iron	Fe	2 or 3
Lead	Pb	2 or 4
Magnesium	Mg	2
Oxygen	O	2
Potassium	K	1
Silver	Ag	1
Sodium	Na	1
Zinc	Zn	2

Radical	Symbol	Valency
Hydroxide	$-OH$	1
Chloride	$-Cl$	1
Nitrate	$-NO_3$	1
Sulphate	$-SO_4$	2
Phosphate	$-PO_4$	3
Carbonate	$-CO_3$	2
Sulphite	$-SO_3$	2
Sulphide	$-S$	2
Nitrite	$-NO_2$	1
Hydrogencarbonate (bicarbonate)	$-HCO_3$	1
Hydrogensulphate (bisulphate)	$-HSO_4$	1
Hydrogensulphite (bisulphite)	$-HSO_3$	1
Oxide	O	2
Ammonium	NH_4-	1

Writing formulae

This is quite a simple operation provided the symbols and valencies are known. Write the valency above the respective symbols of the elements or radicals in the compound and 'cross multiply'. This is best understood by studying the following examples.

Sodium chloride	$\overset{1}{Na}\ \overset{1}{Cl}$	$\longrightarrow NaCl$
Sodium sulphate	$\overset{1}{Na}\ \overset{2}{SO_4}$	$\longrightarrow Na_2SO_4$
Sodium phosphate	$\overset{1}{Na}\ \overset{3}{PO_4}$	$\longrightarrow Na_3PO_4$
Copper (II) nitrate	$\overset{2}{Cu}\ \overset{1}{NO_3}$	$\longrightarrow Cu(NO_3)_2$
Ammonium chloride	$\overset{1}{NH_4}\ \overset{1}{Cl}$	$\longrightarrow NH_4Cl$
Ammonium sulphate	$\overset{1}{NH_4}\ \overset{2}{SO_4}$	$\longrightarrow (NH_4)_2SO_4$
Iron (III) hydroxide	$\overset{3}{Fe}\ \overset{1}{OH}$	$\longrightarrow Fe(OH)_3$
Calcium hydrogen carbonate	$\overset{2}{Ca}\ \overset{1}{HCO_3}$	$\longrightarrow Ca(HCO_3)_2$
Copper (II) oxide	$\overset{2}{Cu}\ \overset{2}{O}$	$\longrightarrow Cu_2O_2$

This last example is written as CuO, this being the simplest formula. Cu_2O_2 means CuO+CuO or two molecules of copper (II) oxide.

Chemical equations

These represent in terms of symbols a chemical change.
E.g.
sodium hydroxide + hydrochloric acid = sodium chloride + water.
(1) $NaOH + HCl = NaCl + H_2O$

When calcium carbonate is heated calcium oxide and carbon dioxide are formed.
(2) $CaCO_3 = CaO + CO_2$

Balancing equations

In both the examples given above the left-hand side equals the right-hand side; the equations are said to be **balanced.** An unbalanced equation might be:

(3) $NaOH + H_2SO_4 = Na_2SO_4 + H_2O$.

In this case the left-hand side consists of **1 atom of sodium 3 atoms of hydrogen**, 5 atoms of oxygen and 1 atom of sulphur, while the right-hand side has **2 atoms of sodium**, **2 atoms of hydrogen**, 5 atoms of oxygen and 1 atom of sulphur. To balance the equation the appropriate numbers should be written (after thought) in front of the formulae thus:

(4) $\mathbf{2}NaOH + H_2SO_4 = Na_2SO_4 + \mathbf{2}H_2O$

Information from equations

Equations state:

1. the reactants and resultants
2. the physical state of the resultants and reactants according to the following convention:
   (s) – solid, (g) – gas, (aq) – aqueous solution, (l) – liquid.
   E.g. $Na(s) + H_2O(l) \longrightarrow NaOH(aq) + H_2(g)$.
   Balanced, this becomes:
   $2Na(s) + 2H_2O(l) \longrightarrow 2NaOH(aq) + H_2(g)$.
3. the ratios of the reacting compounds
   e.g. $\mathbf{2}NaOH + H_2SO_4 \longrightarrow Na_2SO_4 + \mathbf{2}H_2O$
   2 mols + 1 mol ⟶ 1 mol + **2** mols

Equations do not state the conditions of the reaction, e.g. whether heat is required, the concentration of the reactants, whether the reaction is slow or fast, etc.

Ionic equations

Sometimes it is possible to write down equations in relation to ions (electrically charged particles, see chapter 5) and not molecules.
E.g. When hydrogen gas is passed over heated copper (II) oxide, copper is formed together with water.

The molecular equation for this reaction is:

$$CuO + H_2 \longrightarrow Cu + H_2O$$

This equation indicates that copper ions are changed into copper atoms while 'free' hydrogen (H_2) is converted into 'combined' hydrogen (H_2O). As a result the equation can be written:

$$Cu^{2+} + 2e \longrightarrow Cu$$
$$H_2 \quad - 2e \longrightarrow 2H^+$$

Adding $Cu^{2+} + H_2 \longrightarrow Cu + 2H^+$ – ionic equation

Note

(1) that the charges on each side are equal;

(2) that there is no net gain or loss of electrons.

E.g. In the equation $H_2 - 2e \longrightarrow 2H^+$ it would appear at first sight that there are **two negative** charges on the left-hand side and **two positives** on the right-hand side. In fact, on the left-hand side two negative particles are given up resulting in a net charge of 2+.

E.g. $H_2 - (-2e)$, net charge 2+, because electrons are negatively charged in their own right.

There are numerous reactions which can be written ionically, particularly where oxidation and reduction processes are involved (see later chapters). Another example is the reaction between copper sulphate solution and zinc. The molecular equation is:

$$Zn + CuSO_4 \longrightarrow ZnSO_4 + Cu \qquad \text{or ionically}$$

$$Zn + 2e \longrightarrow Zn^{2+}$$

$$Cu^{2+} - 2e \longrightarrow Cu$$

$$Cu^{2+} + Zn \longrightarrow Cu + Zn^{2+} \qquad \text{— ionic equation}$$

It is not always necessary to write ionic equations in two stages and then add. However, this method does enable students to see quite clearly how the final equation is formulated.

Key terms

Atom The smallest particle of an element which can exist.

Molecule The smallest particle of an element or compound that can normally exist.

Radical A group of atoms which can exist in compounds, but do not and cannot exist in their own right, e.g. sulphate radical $-SO_4$, ammonium radical NH_4-.

Valency The valency of an element is the number of hydrogen atoms which will combine with or be displaced by one atom of another element, i.e. the combining capacity of an element.

Chemical equation A representation of a chemical change in terms of symbols.

Chapter 4
Characteristics of Chemical Changes

Physical and chemical changes

When water freezes or boils it is quite easy to obtain the original water by melting the ice or condensing the steam formed. Ice, water and steam are three different states of one compound, and in converting one to the other there is simply a **change of state**. Changes such as the ones mentioned are easily reversible and are called **physical changes**. Other examples of physical changes are wax (or other substances) melting, ammonium chloride or iodine subliming, and liquids boiling. In all these cases it is quite easy to reverse the reaction.

If coal were burned and all the products of burning were collected together, it would be virtually impossible to convert the products of burning coal back into coal. The change is not easily reversible and is a good example of a **chemical change**. In this particular reaction new substances are formed, e.g. gases, coke etc., and heat and light are given out. Magnesium ribbon when burned in air forms magnesium oxide powder and does so with the emission of a blinding light, and a considerable amount of heat. This is another example of a chemical change. Below are summarised the main differences between physical and chemical changes.

	Physical change	Chemical change
(i)	No new substance formed	New substance formed
(ii)	Easily reversible	Usually difficult and often impossible to reverse
(iii)	No great energy change involved	Considerable energy change involved
(iv)	No change in weight	A change or redistribution of weight

Laws and concepts relating to chemical changes

Laws which relate to combination by weight provide evidence for Dalton's atomic theory which postulates:

1. matter is made up of small indivisible particles called atoms;
2. atoms can neither be created nor destroyed;

3. the atoms of any one element are all alike, and different from the atoms of any other element;
4. chemical reactions take place between small, whole numbers of atoms.

1. Law of conservation of mass

Matter can neither be created nor destroyed during the course of a chemical reaction.

This law provides evidence for the second point of the theory and can be illustrated (not proved) in the following way.

Some dilute hydrochloric acid is placed in a conical flask, and an ignition tube containing silver nitrate solution is suspended by a cotton thread around the top of the tube, held in place by a cork in the mouth of the flask. The apparatus is then weighed. The cotton is released emptying the silver nitrate into the hydrochloric acid. The apparatus is reweighed and it is found that no change in weight has occurred.

Note A chemical reaction has taken place because a white precipitate of silver chloride is seen to have been formed.

$$AgNO_3 + HCl \rightarrow AgCl \downarrow + HNO_3$$

2. Laws of constant composition

The composition of a chemical compound is always the same irrespective of how it is made.

This law is evidence for the third point of the atomic theory and can be illustrated thus.

Prepare copper (II) oxide (CuO) in **two** different ways.

1. Heat copper (II) carbonate ($CuCO_3$) until the weight of the copper (II) oxide residue is constant.

$$CuCO_3 = CuO + CO_2 \uparrow$$

2. Heat copper (II) nitrate ($Cu(NO_3)_2$) until the weight of the copper (II) oxide residue is constant.

$$2Cu(NO_3)_2 = 2CuO + 4NO_2 \uparrow + O_2 \uparrow$$

The copper oxide samples from each experiment are then reduced separately in the apparatus shown in figure 12, exactly the same weight of oxide being reduced in each case.

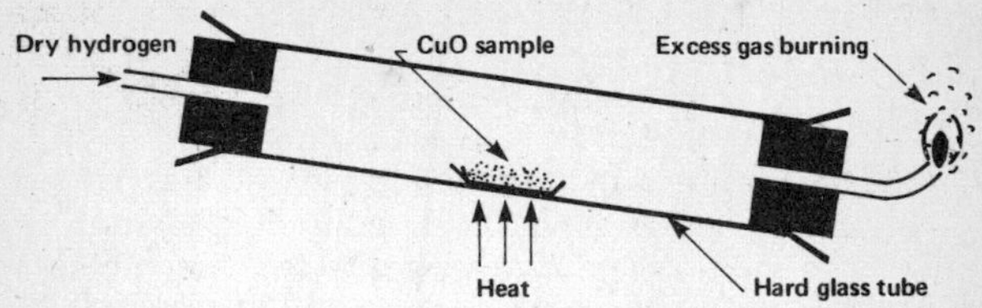

Figure 12. Oxide reduction apparatus

Notes on the oxide reduction experiment

1. The excess hydrogen which is burning at the end of the tube must be tested before it is lit.
2. The combustion tube must be tilted as shown to prevent water which has condensed at the end of the tube from running back and cracking the very hot central portion of the tube.
3. The residue must be heated and cooled with hydrogen gas still passing over in order to prevent the possibility of back oxidation.
4. The heating and cooling must be continued until a constant weight is obtained.

$CuO + H_2 = Cu + H_2O$ (molecular equation).

$Cu^{2+} + H_2 = Cu + 2H^+$ (ionic equation).

Equal weights of both samples will give equal weights of copper, thus illustrating the law.

3. Law of multiple proportions

If two elements A and B combine together to form more than one compound, then the different weights of A that combine with a fixed weight of B are in a simple whole number ratio.

This law provides evidence for the fourth point of the theory. E.g. two oxides of copper:

Copper (I) oxide Cu_2O and copper (II) oxide CuO.

The ratio of copper to oxygen is 2:1 and 1:1 respectively. Similarly in lead (II) oxide PbO and lead (IV) oxide PbO_2 the ratio of lead to oxygen is 1:1 and 1:2 respectively.

To illustrate this law the apparatus shown in figure 12 can again be used, together with the same precautions that applied when illustrating the law of constant composition. Known weights of lead (II) oxide (litharge PbO), lead (IV) oxide (PbO_2) and red lead oxide (Pb_3O_4) are reduced to lead by heating in a stream of hydrogen.

$$PbO_2 + 2H_2 = Pb + 2H_2O \text{ (molecular equation).}$$

$$Pb^{4+} + 2H_2 = Pb + 4H^+ \text{ (ionic equation).}$$

On cooling each sample of lead is weighed and the process of reduction, cooling, and weighing repeated until constant weights are obtained in each sample. The weight of lead combining with 1g of oxygen in each oxide is then calculated. It is found that the different weights of lead combining with 1g of oxygen are in a simple ratio.

E.g. PbO $\quad PbO_2 \quad Pb_3O_4$

$$4 : 2 : 3$$

The three laws all relate to combination by weight. There are however other important laws which are relevant to volumes.

Gay Lussac's law of gaseous combining volumes

When gases react they do so in volumes which are simply related to each other and to that of the product if gaseous. (All measurements must be made under the same conditions of temperature and pressure.)

Important volume compositions

1. Steam

2 volumes of steam contain 2 volumes of hydrogen and 1 volume of oxygen.

$$\underset{\text{2 vols}}{2H_2} + \underset{\text{1 vol}}{O_2} = \underset{\text{2 vols}}{2H_2O}$$

2. Hydrogen chloride

2 volumes of hydrogen chloride contain 1 volume of hydrogen and 1 volume of chloride.

$$\underset{\text{1 vol}}{H_2} + \underset{\text{1 vol}}{Cl_2} = \underset{\text{2 vols}}{2HCl}$$

3. Ammonia

2 volumes of ammonia contain 1 volume of nitrogen and 3 volumes of hydrogen.

$$\underset{\text{1 vol}}{N_2} + \underset{\text{3 vols}}{3H_2} = \underset{\text{2 vols}}{2NH_3}$$

Avogadro's law

Equal volumes of all gases under the same conditions of temperature and pressure contain the same number of molecules.

E.g. If we had two identical containers and put hydrogen in one and oxygen in the other, under the same conditions of temperature and pressure, each container would hold the same number of molecules of the respective gases.

If we took the molecular weight of hydrogen in grams, i.e. 2g or 32g of oxygen ($O_2 = 32$) and measured the respective volumes at s.t.p., then it would be found to be 22.4 litres (dm^3) in each case. I.e. 1 molecular weight in grams, which is called a **mole**, of any gas at s.t.p. occupies 22.4 litres (called the **molar volume of a gas**). Moreover, 1 **mole** of a gas or 1 **mole** of any element or compound will contain the same number of particles = 6.02×10^{23}. This is known as the **Avogadro number or constant.**

The term **mole** now replaces terms such as gram atom (atomic weight in grams), g. molecule (molecular weight in grams), g. formula (formula weight in grams), g. ion, etc.

Examples

1. 1 mole of chlorine atoms is 71 grams, i.e.
   $Cl_2 = Cl + Cl = 35.5 + 35.5$.
   71 grams used to be called a gram molecule but are now known as 1 mole.
2. The compound sodium carbonate has the formula Na_2CO_3.
   1 mole of sodium carbonate = 106g
   i.e. 2[Na] (46) + C(12) + 3[O](48) = 106
   This 106 used to be called the **formula weight** or **molecular weight**, while 106g used to be called the **gram formula weight** or **gram molecular weight**; now however, 106 grams of sodium carbonate (Na_2CO_3) is referred to as **1 mole**.
3. **1 mole** of the element copper would be 64 grams of copper. This used to be called a **gram atom**.

Note 71g of chlorine, 106g of sodium carbonate, and 64g of copper would each contain the Avogadro number of molecules, i.e. $6.023 \times 10^{23}\ mol^{-1}$.

Molar solutions

A molar solution is one that contains 1 mole of a substance in 1 litre (dm^3) of a solution.

E.g. to make a molar solution of sodium hydroxide (NaOH), 1 mole of the substance (40g) should be weighed out accurately, dissolved in distilled water and the volume made up to exactly 1 litre (dm^3). This solution would contain 1 mole of sodium hydroxide and would be referred to as M.NaOH (see Modern nomenclature, below). 25 cm^3 of the solution would contain .025 moles of solute. Often it is necessary to make solutions which are weaker than 1 M (1 mol/l), e.g. 0.1 M (0.1 mol/l) or 0.05 M (0.05 mol/l). This is quite simple because a 0.1 M (0.1 mol/l) solution of sodium hydroxide would contain 0.1 of a mole, i.e. 4.0g/l (dm^3) of sodium hydroxide whereas a 0.05 M (0.05 mol/l) solution would contain 0.05 of a mole of sodium hydroxide = 40 × .05 = 2.0g/l (dm^3). By definition, if the concentration of a solution is expressed in **molar** terms then the **number of particles** in the solution is being considered.

Modern nomenclature

In recent years there has been a change in the nomenclature used, relevant parts of which are summarised below.

Atomic weights

These **were** written as Na = 23, Cl = 35.5, etc. **Now** they are written thus:

$$A_r(Na) = 23 \quad A_r(Cl) = 35.5$$

Molecular weights

The molecular weight of the compound tetrachloromethane (carbon tetrachloride) CCl_4 **was** written $CCl_4 = 154$. **Now** this is written $M_r(CCl_4) = 154$. Similarly the molecular weight of sodium carbonate $M_r\ (Na_2CO_3) = 106$.

Molar mass

G. atom, g. molecule, g. ion, g. formula weight are now referred to as the **molar mass** of the entity, e.g. 1g atom of nitrogen would be written as the molar mass of nitrogen M(N) (a mole of nitrogen), M(Cl) = 35.5 g/mol, $M(CH_4)$ = 16g/mol, $M(MnO_4{}^{2-})$ = 119 g/mol, M(NaCl) = 58.5 g/mol.

These changes are best illustrated by transposing a question into the new terminology. E.g. how many g formulae of sodium hydroxide are contained in 250 cm^3 of 2.0 M (2.0 mol/l) sodium hydroxide solution?

Note The symbol M in the new terminology is being used in relation to molar mass. It therefore cannot now be used for concentrations of solutions. The new nomenclature for 2.0 M NaOH is 2.0 mol/l NaOH or 2.0 mol/dm^3. The question would now read:

How many moles of sodium hydroxide are contained in 250 cm^3 of 2.0 mol/1 sodium hydroxide solution?

Miscellaneous calculations

1. Calculate the number of moles in:

    (a) 46 grams of sodium (b) 16 grams of sulphur.
    $A_r(Na) = 23$, $A_r(S) = 32$.

    Answer

    (a) There are 23g of sodium in 1 mole

    $$1\text{g of sodium} = \frac{1}{23}\text{ mole}$$

    $$\therefore \text{ in 46g of sodium there are } \frac{1}{23} \times 46 \text{ moles}$$

    $$= 2 \text{ moles of sodium.}$$

    (b) Using the same reasoning it can be calculated that there is 1/2 mole of sulphur in 16g of the element.

2. 0.05 mole of sodium carbonate is treated with excess hydrochloric acid solution. 1.12 litres (dm^3) of carbon dioxide at s.t.p. are produced. Calculate what fraction of a mole of carbon dioxide is contained in 1.12 litres (dm^3).
    (molar volume = 22.4 1 at s.t.p.) $M_r(CO_2) = 44$.

    Answer

    22.4 litres of CO_2 are obtained from 1 mole of the gas.

    $$1 \text{ litre of } CO_2 \text{ would come from } \frac{1}{22.4} \text{ moles}$$

    $$\therefore 1.12 \text{ litres of } CO_2 \text{ would come from } \frac{1}{22.4} \times 1.12 = \frac{1}{20} \text{ mole.}$$

3. What volume of 1.0 mol/l hydrochloric acid would be required to just neutralise 10 cm^3 of 1.0 mol/l (molar) sodium carbonate solution?

Answer

$2HCl + Na_2CO_3 = 2NaCl + H_2O + CO_2 \uparrow$

From the equation:

10 cm^3 of 1.0 mol/l sodium carbonate would need **2** × 10 cm^3 of 1.0 mol/l hydrochloric acid solution.

Answer = 20 cm^3 1.0 mol/l HCl.

Note that the ratio HCl : Na_2CO_3 is 2 : 1 in the equation.

Energy changes

When water is added dropwise to quicklime, or concentrated sulphuric acid is poured carefully into cold water, there is in each case a large **increase in temperature**, and the reactions are said to be **exothermic**.

If ammonium chloride is dissolved in water there is a sharp **decrease in temperature** or a net loss of heat, and the reaction is said to be **endothermic**.

Chemical changes or reactions are usually associated with energy changes, these being for the most part much greater than those associated with ordinary physical changes where the net loss or gain of energy is usually small.

An **exothermic reaction** can be defined as one in which energy is **given out**.

An **endothermic reaction** is one in which energy is **taken in**.

When a fuel burns heat energy is given out, and this type of reaction is doubtless in this day and age the most important of the exothermic reactions. When there is a change of energy in a reaction it can be indicated at the end of the equation by the symbol $\triangle H$. If $\triangle H$ is negative the reaction is exothermic while $\triangle H$ is positive for a reaction which is endothermic thus:

$$2C(s) + O_2(g) = 2CO(g) + \text{heat}$$

or $2C(s) + O_2(g) = 2CO(g) \quad \triangle H = -221\text{k.J/mol}$

The equation tells us that if 24g of carbon were reacted with 32g of oxygen then 56g of carbon monoxide would be produced and 221k.J of energy would be given out. The letters in brackets indicate the physical state of the reactants, s=solid, g=gas, etc.

$$C(s) + H_2O(g) = CO(g) + H_2(g) - \text{heat}$$

or $C(s) + H_2O(g) = CO(g) + H_2(g), \ \triangle H = +131\text{k.J/mol}$

This is an endothermic reaction.

Common examples of exothermic and endothermic reactions

Exothermic reactions ($\triangle H$ is negative)

$$C + O_2 = CO_2$$
$$2H_2 + O_2 = 2H_2O$$
$$H_2 + Cl_2 = 2HCl$$
$$2Mg + O_2 = 2MgO$$
$$CaO + H_2O = Ca(OH)_2$$

Endothermic reactions ($\triangle H$ is positive)

$$H_2O + C = H_2 + CO$$
$$H_2 + I_2 = 2HI$$
$$C + 2S = CS_2$$
$$CaO + 3C = CaC_2 + CO$$

Respiration and **photosynthesis** are examples in nature of exothermic and endothermic reactions respectively.

During **respiration** living organisms release energy from food material by means of a series of chemical reactions. Among these reactions is one which involves the oxidation of carbohydrates to form carbon dioxide and water, with the corresponding energy release.

$$C_6H_{12}O_6 + 6O_2 = 6CO_2 + 6H_2O, \quad \triangle H \quad \text{is negative}$$

Oxygen is produced in the endothermic reaction called **photosynthesis** by green plants building up carbohydrates from water and carbon dioxide. The energy required for this reaction is obtained from sunlight.

$$6CO_2 + 6H_2O = C_6H_{12}O_6 + 6O_2, \quad \triangle H \quad \text{is positive}$$

The two reactions are part of the carbon cycle.

Energy changes in a chemical reaction may be represented by diagrams.

In an exothermic reaction, less energy is associated with the products than with the reactants, and the difference is the amount of energy which is evolved in the form of heat.

However, in an endothermic reaction more energy is associated with the products than with the reactants, and it is the difference between these two values which is the amount of heat absorbed in the reaction. A **compound** which is formed from its elements

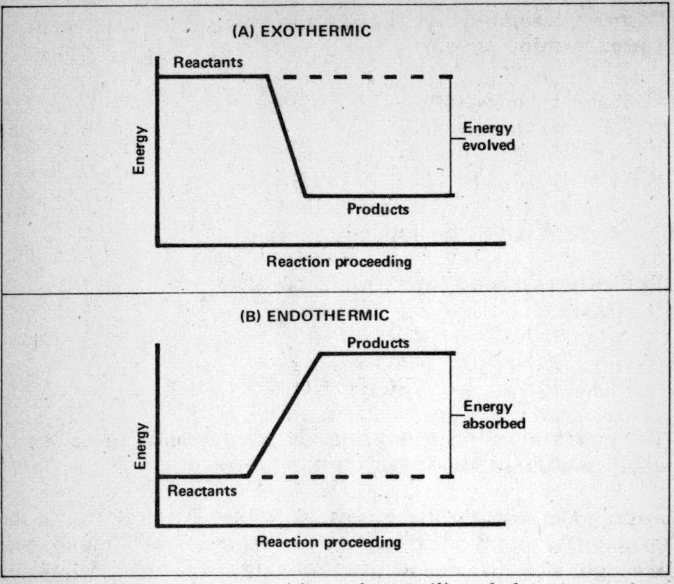

Figure 13. Energy changes in (a) exothermic (b) endothermic reactions

by a reaction in which there is an absorption of heat is called an **endothermic compound**. Endothermic compounds, e.g. acetylene, ozone etc. tend to be unstable and decompose readily whereas exothermic compounds, e.g. hydrogen chloride and water, are usually very stable, and not easily decomposed. Since decomposition involves supplying the same quantity of heat given out when the compound was formed, then this can only be achieved by employing high temperatures or electrolysis.

The rate of a chemical change

Many chemical changes take place almost instantaneously, e.g. the immediate appearance of a white precipitate when silver nitrate solution is added to an aqueous solution of sodium chloride.

$$AgNO_3(aq) + NaCl(aq) = AgCl(s)\downarrow + NaNO_3(aq)$$

Other changes take place at slower rates, e.g. the dissolution of marble chips when acted upon by hydrochloric acid.

$$CaCO_3(s) + 2HCl(aq) = CaCl_2(aq) + H_2O(l) + CO_2(g)$$

The rusting of iron under normal atmospheric conditions is an example of an even slower reaction.

Activation energy

Many reactions need energy to start them, and then proceed using the energy evolved in the reaction. A certain minimum energy has to be reached in many reactions in order that they can proceed. This minimum or initiation energy required is called the **activation energy.**

Factors affecting the rate of a chemical change

The speed at which a reaction takes place depends upon a number of different factors:

1. the reactants;
2. the concentration of the reactants or, if the reaction is involved with gases, their respective pressures;
3. catalysts;
4. the physical state of the reactants;
5. temperature;
6. light.

1. The reactants

Reactions between ionic compounds are fast compared with those between covalent compounds (see page 60). Quite a large proportion of reactions between ionic compounds are almost instantaneous or take a short time to reach completion when compared with reactions between covalent compounds which quite often take hours for any appreciable changes to take place. E.g. If an **alkali** reacts with an **acid** a **salt** is formed by the process of neutralisation. The reaction requires no heat and is completed in a very short time.

$$NaOH(aq) + HCl(aq) = NaCl(aq) + H_2O(l)$$

When an **ester** is prepared by the action of an **alcohol** with an **acid,** the reactants have to be refluxed in the presence of concentrated sulphuric acid for some hours for any major conversion of the reactants to be achieved.

$$C_2H_5OH(l) + CH_3COOH(l) \rightleftharpoons CH_3COOC_2H_5(l) + H_2O(l)$$

This is a **reversible reaction** (see page 47).

2. The concentration of the reactants

For a reaction to take place there must be collisions between the molecules of the respective reactants. It follows that the more molecules present the greater will be the incidence of collision, and hence the faster the reaction will proceed. A 10 mol/l (10M) solution of a substance would contain 10 moles of the substance contained in 1 litre (dm^3). However in a 1.0 mol/l (1M) solution of the same substance there would be 1 mole contained in 1 litre (dm^3). The first solution contains ten times the number of molecules compared with the second solution or is said to be ten times more concentrated. Any reaction involving these solutions would reach completion far more quickly with the 10 mol/l (10M) solution than the 1.0 mol/l (1M) solution. The effect of concentration on the rate of a reaction can be demonstrated by letting identical samples of calcium carbonate react with a series of hydrochloric acid solutions ranging from weak to concentrated. The respective times of completion of each reaction should be noted, and it will be found that the more concentrated the acid solution the faster the reaction rate.

Molecular concentration can be increased in the case of gaseous reactions by increasing the pressure. Increasing pressure on gases decreases their volume and increases the molecular concentration (see figure 14). The reverse is also true.

3. Catalysts

When potassium chlorate is **strongly** heated in a test tube it breaks down slowly into potassium chloride, and oxygen gas is evolved.

$$2KClO_3 = 2KCl + 3O_2$$

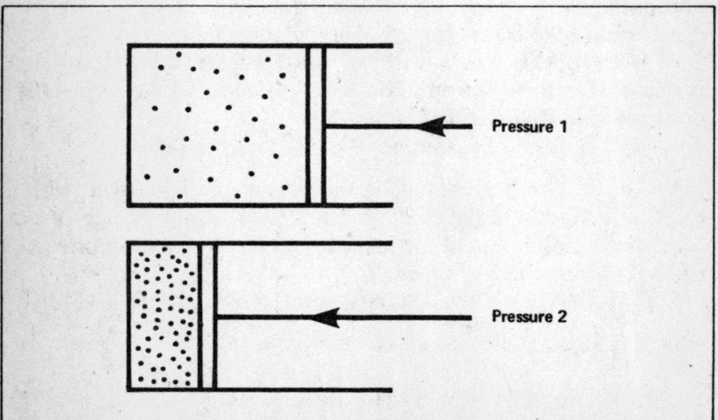

Figure 14. Effect of pressure on gases

When a mixture of potassium chlorate and manganese (IV) oxide is gently heated oxygen is evolved at a much faster rate, and the overall reaction is faster. The manganese (IV) oxide is unchanged in mass and chemical composition at the end of the experiment. It is obvious that the manganese (IV) oxide has in some way helped the second reaction to proceed at a faster rate, and at a lower temperature. It is said to have acted as a **catalyst**.

A catalyst is a substance which affects the rate of a chemical reaction without itself being changed in mass or chemical composition during the course of the reaction.

Catalysts are not used up in chemical reactions, although they may take part in it and their physical state may be altered. They do not catalyse **any** reaction, but only those of a particular type suited to the catalyst. They may increase the rate of a reaction, when they are referred to as **positive catalysts**. Catalysts can also decrease the rate of a chemical reaction and are thus called **negative catalysts or inhibitors**, e.g. specially prepared liquids are added to water in order to decrease corrosion in pipes, radiators, etc.

Reactions in which catalysts are used include:

(i) The contact process for the manufacture of sulphuric acid, platinised asbestos or more recently vanadium pentoxide being the catalyst employed.

(ii) The Haber process in which hydrogen and nitrogen are synthesised to make ammonia gas. The catalyst in this case is made up of a large proportion of iron.

(iii) The addition of copper will increase the rate of the reaction between dilute sulphuric or hydrochloric acids and zinc.

(iv) In the processing of crude oil catalytic cracking is employed using aluminium oxide as the catalyst.

Catalysts are being used more and more in the chemical industry and many reactions would be uneconomic unless suitable catalysts were used to increase the reaction rates. They are generally required only in relatively small quantities.

4. The physical state of the reactants

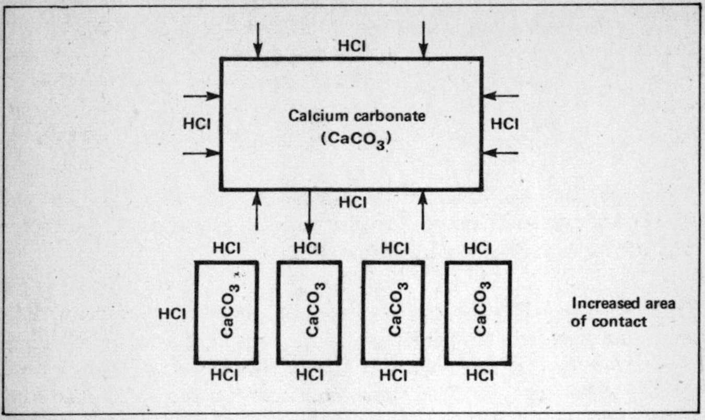

Figure 15. The physical state of reactants

When dilute hydrochloric acid reacts with calcium carbonate, the more calcium carbonate in contact with the acid the greater will be the reaction rate. A single piece of calcium carbonate would not react as fast with the acid as the same piece broken down into a number of smaller pieces thereby increasing the area of contact with the acid. Figure 15 shows that there is a greater area of contact in one case than the other resulting in a faster reaction rate.

5. Temperature

The higher the temperature of the reactants the faster the reaction occurs. Heat energy increases the kinetic movement of molecules hence increasing the incidence of collision between the molecules and the incidence of product formation by reaching or exceeding the activation energy for the reaction to proceed.

6. Light

Light increases the rate of some reactions. If the rate of a reaction is affected by light it is called a **photochemical** or **photo-sensitive** reaction.

E.g. Silver salts, particularly silver chloride, are decomposed by light forming silver.

A mixture of hydrogen and chlorine will not react together, and will remain in a stable state in the dark. However, when exposed to direct sunlight they react with explosive violence to form hydrogen chloride gas.

$$H_2(g) + Cl_2(g) = 2HCl(g)$$

In this particular reaction the light breaks the diatomic chlorine atom Cl_2, which in turn reacts with hydrogen H_2 to form HCl, freeing a hydrogen atom for further reaction and so on.

$$Cl_2 \xrightarrow{\text{light}} Cl + Cl$$
$$Cl + H_2 \longrightarrow HCl + H$$
$$H + Cl_2 \longrightarrow HCl + Cl \text{ etc.}$$

Reversible reactions

Whilst some reactions go to completion exactly as indicated in the chemical equation, other reactions do not. Ionic reactions involving the precipitation of a solid are usually irreversible, as are reactions involving the liberation of a gas.

E.g. $Ag^+ + Cl^- = AgCl\downarrow$

$Mg + 2H^+ = Mg^{2+} + H_2\uparrow$

However, many reactions are reversible and can go in either direction according to the conditions under which the reaction is carried out. Consider the reaction between steam and iron.

$$3Fe(s) + 4H_2O(g) \rightleftharpoons Fe_3O_4(s) + 4H_2(g)\uparrow$$

If iron and steam are heated in a closed vessel so that none of the substances are removed, a point is reached at which the concentrations of all four substances do not alter, unless the conditions are changed. The reaction is then said to be 'balanced' and the system is in equilibrium. The equilibrium is **dynamic** not static, i.e. the substances are still reacting, but the rate of reaction from left to right (the forward reaction) is exactly equal to the rate of the reaction from right to left (the backward reaction). Thus the concentrations of all four substances remain constant. Reversible reactions are common and include:

1. $CaCO_3(s) \rightleftharpoons CaO(s) + CO_2(g)$
2. $NH_4Cl(s) \rightleftharpoons NH_3(g) + HCl(g)$
3. $H_2O \rightleftharpoons H^+ + OH^-$
4. $CH_3COOH \rightleftharpoons CH_3COO^- + H^+$

The first two reactions are examples of **thermal dissociation** and the last two reactions examples of **ionic dissociation**. Thermal dissociation should not be confused with thermal decomposition; both changes are brought about by heat but thermal dissociation is reversible and the products recombine on cooling, whereas thermal decomposition is irreversible.

Effect of conditions on reversible reactions

Two factors must be considered:

1. effect on the rate of both forward and backward reactions;
2. effect on the equilibrium position.

Concentration

$$A + B \rightleftharpoons C + D$$

If the concentration of either A or B is increased the rate of the forward reaction increases with the consequent increase in concentration of C and D until equilibrium is again established.

Temperature

Increase in temperature causes an increase in the rates of both forward and backward reactions to the same extent. Thus the equilibrium position is unchanged, but equilibrium is reached more quickly.

$$A + B \rightleftharpoons C + D + \text{heat}$$

If the forward reaction is exothermic, when the temperature is raised the system adjusts itself in order to try to reduce the temperature again. This can only be achieved by a change of some C and D into A and B with the result that the position of equilibrium alters in favour of the backward reaction, absorbing some of the heat. The forward reaction is therefore favoured by low temperatures.

Pressure

Where there is no change in the total volume of the system when either the forward or backward reaction occurs, increase in pressure merely increases the rate of the forward reaction and backward reactions to the same extent, and the position of equilibrium is unaltered. However, consider (i) $A + B \rightleftharpoons C$ (ii) $A \rightleftharpoons B + C$. Increase in pressure causes the reaction to move so that the total volume of gases is decreased. Therefore in (i) more C is formed and in (ii) more A is formed, i.e. backward reaction is favoured.

Practical examples

1. Manufacture of ammonia (Haber process)

$$\underset{\text{1 vol}}{N_2} + \underset{\text{3 vols}}{3H_2} \rightleftharpoons \underset{\text{2 vols}}{2NH_3} + \text{heat}$$

Since the forward reaction is exothermic, it will be favoured by a low temperature. The reason for this is that if the temperature is raised the system adjusts itself to try to reduce the temperature again; this can only be achieved by a change of some of the ammonia into nitrogen and hydrogen, and thus some of the heat would be absorbed.

As there is a decrease in volume in the forward reaction it is favoured by a high pressure. If the pressure is raised, the system tends to adjust itself to try to reduce the pressure again, and this is achieved by converting more nitrogen and hydrogen to ammonia in order to reduce the volume of the system.

In the Haber process, a high pressure (200 atmosphere) and a moderately high temperature are used together with an iron catalyst. Although a high temperature does not favour the formation of ammonia, equilibrium is reached more quickly. A low yield of ammonia (10 per cent) produced quickly is economically preferable to a higher yield produced slowly.

2. Manufacture of nitric acid

The essential reactions are (a) the oxidation of ammonia to nitrogen monoxide, (b) its subsequent conversion to nitrogen dioxide and (c) the absorption of nitrogen dioxide in water.

(a) $4NH_3 + 5O_2 = 4NO + 6H_2O + \text{heat}$
4 vols 5 vols 4 vols 6 vols

(b) $2NO + O_2 \rightleftharpoons 2NO_2 + \text{heat}$
2 vols 1 vol 2 vols

(c) $2NO_2 + H_2O = HNO_3 + HNO_2$
2 vols 1 vol 1 vol 1 vol

The forward reaction in (a) is favoured by low temperatures since it is exothermic. As there is a slight volume change in the reaction, it is not greatly affected by pressure. The forward reaction in (b) is favoured by low temperature and high pressures since the reaction is accompanied by a decrease in volume.

In practice (a) is carried out using a platinum catalyst at 700–800°C under normal pressures, the temperature being maintained by the heat of the reaction. The nitrogen monoxide is then mixed with air, and cooled to about 150°C to ensure a good yield of nitrogen dioxide under normal pressures.

Le Chatelier's principle
If a system in physical or chemical equilibrium has that equilibrium disturbed, i.e. a change of temperature, pressure or concentration takes place, the system will tend to change in order to restore that equilibrium, i.e. the reaction will follow the path of least resistance.

In the case of important industrial reactions such as the ones listed above, very careful consideration must be given to changing factors in relation to Le Chatelier's principle.

Key terms

Physical change One in which *no* major change in the properties of a substance occurs, e.g. water turning to steam. Such a change is easily reversed.
Chemical change One in which a new substance is formed. Such a change is often associated with heat and light; it is difficult and often impossible to reverse.
Law of conservation of mass Matter cannot be created nor destroyed during the course of a chemical reaction.
Law of constant composition (definite proportions) A given pure chemical compound always contains the same elements combined in the same proportions no matter how it is made.
Law of multiple proportions If two elements combine to form more than one compound, then the weights of one which combine with the fixed weight of the other are in a simple whole number ratio.
Gay Lussac's law of volumes When gases react they do so in volumes which bear a simple ratio to each other, and to that of the product if gaseous.
Avogadro's law Equal volumes of all gases under the same conditions of temperature and pressure contain the same number of molecules.
Molar solution One which contains 1 mole of a substance in 1 litre (dm^3) of a solution.
Mole The molecular weight of a substance expressed in grams, containing the Avogadro number of particles.
Exothermic reaction A reaction in which heat energy is given out.
Endothermic reaction A reaction in which heat energy is taken in.

Activation energy The minimum energy required by a reaction in order that it may proceed.
Catalyst A substance which affects the speed of a chemical reaction, but remains unchanged in any way itself.
Le Chatelier's principle A system which is in physical or chemical equilibrium will, if this equilibrium is disturbed, tend to change in order to restore that equilibrium.

Chapter 5
Atomic Structure and Bonding

An **atom** is the smallest part of an element that can take part in a chemical change.

An **element** is a substance which cannot be split up chemically into anything simpler.

The atom is made up of three fundamental particles, the **electron**, the **proton**, and the **neutron**, the last two being found in the **nucleus** whilst the electrons circulate about the nucleus. The simple atom could be likened to the solar system, the sun being the central nucleus while the planets (electrons) circulate around the sun.

In the table below are shown the relative masses and charges of the three particles.

Particles	**Mass**	**Charge**
Electron	1/1840	−1
Proton	1	+1
Neutron	1	0

Atomic number

All atoms are neutral despite the fact that they contain charged particles. It follows that the positives (protons) must be equal to the negatives (electrons), i.e. the protons = the electrons. The number of protons or electrons in any atom is called the **atomic number**. This atomic number is:

1. the charge on the nucleus (the number of protons in the nucleus);
2. the number of electrons circulating around the nucleus.

E.g. An atom of the element sodium has an atomic number of 11; this means that its nucleus contains 11 protons, and it has 11 electrons circulating around the nucleus. Later we shall see that in the Periodic table of the elements the atomic number is of vital significance.

Atomic mass

A fundamental property of the atom is the **atomic mass**, which is the number of protons + the number of neutrons, e.g. an atom of the element sodium has an atomic number of 11, and an atomic mass of 23; it therefore has 11 electrons, 11 protons and 12 neutrons.

Isotopes

It is possible for atoms of the same element to exist having the same atomic number but different atomic masses. Such atoms are called isotopes.

E.g. Hydrogen has three isotopes or three atoms having a different number of neutrons in each of their nuclei.

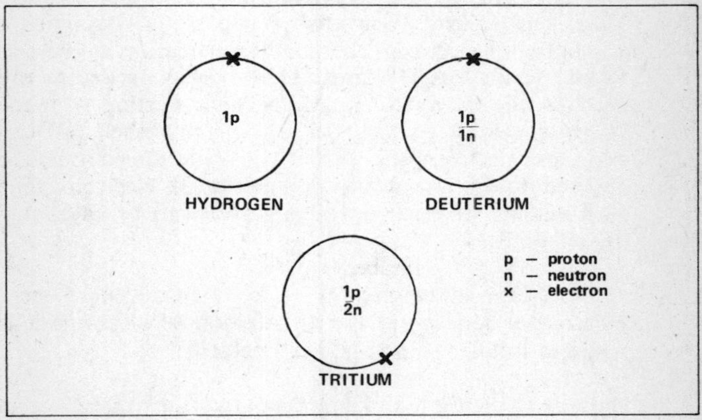

Figure 16. The isotopes of hydrogen

Another example is chlorine which has two isotopes, one of atomic mass 35, and another of atomic mass 37. The two isotopes are written as:

$$^{35}_{17}Cl \quad \text{and} \quad ^{37}_{17}Cl$$

All isotopes are present in constant proportion in elements, and in the case of chlorine a sample of the gas would contain 75% of isotope 35 and 25% of isotope 37, hence the **average atomic mass** is 35.5. This average atomic mass is more commonly called the **atomic weight** of the element and is relative to the atomic mass of the isotope of carbon=12.

The figure 35.5 is calculated as follows:

3 parts (75%) of isotope 35 gives a total	105
1 part (25%) of isotope 37 gives a total	37
Grand total	142

$$\text{Average} = \frac{142}{4} = 35.5$$

Electron distribution

The electrons which revolve around the nucleus do so according to a definite pattern. Groups of electrons maintain definite average distances from the nucleus by forming **shells** of electrons surrounding the nucleus. Each **shell** is capable of containing a definite number of electrons, the number increasing as the distance from the nucleus increases. These shells or **orbits** are designated by the letters K, L, M, N, etc., starting with the shell which is nearest to the nucleus. The first shell (K) can contain up to 2 electrons, the second (L) up to 8 electrons, the third (M) up to 18, the fourth (N) up to 32 electrons. The maximum number of electrons in any shell can be calculated from the relationship:

$$\text{number} = 2S^2$$

where S = number of the shell (K=1, L=2, M=3, N=4, etc.). In calculating or looking at the distribution of electrons it is worthwhile to follow the points listed below.

1. In the first 18 elements (hydrogen to argon) the new electron is added to the outermost shell until the shell is filled, then a new shell started.

2. In the higher-numbered elements there could be two or even three unfilled electron shells.

3. **Eight** electrons temporarily fill each of the shells beyond and including the M shell, and a new shell must be started before more electrons can be placed into the temporarily-filled shell.

4. The golden rule to remember is that there are **never** more than 8 electrons in the outermost shell.

On the above basis it is possible to recognise **four** different structural types of elements:

1. inert elements – all shells filled;

2. simple elements – one unfilled shell;

3. transition elements – two unfilled shells;

4. rare earth elements – three unfilled shells.

On the basis of electronic distribution all the elements have been arranged on a chart called the Periodic table (see figure 26). The vertical columns are called **groups** and all elements in a group have the same electronic structure in their outermost shell e.g. all the Group 1 elements have one electron in the outermost shell, those in Group 2 have two, etc. The transition elements are arranged in sub-groups and all have two outermost electrons with the exception of the copper/silver group which has only one outermost electron.

The horizontal rows of elements are called **periods**. All the elements in a given period have the same number of shells of electrons, e.g. the elements in Period 1 have one shell of electrons, those in Period 2 have two, etc. It is important to note that the last element in each period is an inert element, i.e. one which has eight electrons in its outermost shell.

E.g. Element X has an atomic number of 15. It has an electronic distribution of 2, 8, 5 and will therefore be found in Group 5 Period 3 of the table. From this example it can be seen that it is the atomic number, i.e. the electronic distribution, which dictates the chemical properties of elements and in particular the structure of the outermost shell or orbit. Since each group of elements has the same structure on the outermost shell, then it would be expected that members of a group will show **similar** chemical behaviour; for this reason the groups can be regarded as 'families of elements', and the Periodic table will be found constantly useful as the properties and nature of the elements are explored.

Bonding

Elements in a free or uncombined state make up only a very small part of matter. Most matter is made up of compounds. The atomic arrangement described previously is very useful in showing how elements can combine to form the various compounds. It is important to realise that when a compound is formed or a chemical change takes place, only the electrons are involved, the nuclei of atoms being in no way affected during the formation of a compound.

The structure of the **inert gases** is of great significance in examining the formation of compounds and the types of bonds which result. Apart from helium which has two electrons in its outermost shell, the other noble gases have eight electrons (remember in the Periodic table a group always ends with an inert gas and that there are eight groups), i.e. an octet of electrons in the outer shell. These gases as a result of their electronic structure are very stable, and show very little tendency to react in any way. They form no compounds among themselves or with any other elements except with great difficulty.

All other elements are less stable since they do not possess this octet of electrons on the outer shell, and they do react to form compounds. It is suggested that when active elements combine to form compounds they undergo a rearrangement of their electronic configuration similar to that of an inert element. Such a rearrangement causes the active element to become more stable.

In order to achieve this rearrangement, i.e. the formation of a stable octet of electrons on the outer shell, when two elements combine together one or more electrons may be **transferred** from one atom to another forming an **electrovalent bond**; or one or more pairs of electrons may be **shared** between two atoms forming a **covalent** bond.

Electrovalent or ionic bond formation

An atom of sodium has one electron on its outside shell, and eight electrons in its next innermost shell. Sodium has only to lose this single electron to have an electronic arrangement of eight electrons on the outside shell comparable with the inert gases. Chlorine on the other hand has seven electrons on its outermost shell, and requires only one electron to attain its octet. In the formation of sodium chloride (common salt) sodium **gives** chlorine one electron and the compound sodium chloride is formed.

Common ionic compounds

1. Copper (II) sulphate: $CuSO_4$
2. Potassium nitrate: KNO_3
3. Ammonium chloride: NH_4Cl
4. Sodium nitrate: $NaNO_3$
5. Copper (II) nitrate: $Cu(NO_3)_2$

(2–4: almost all potassium, ammonium and sodium salts)

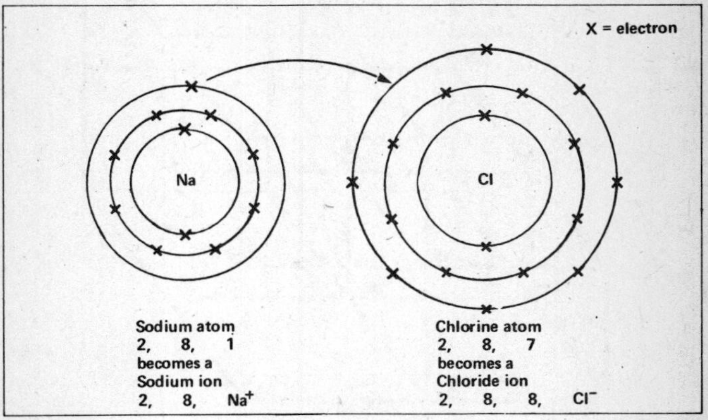

Figure 17. Electron transfer to form an ionic bond

As a result of the above interaction sodium now has one excess positive charge, i.e. the protons do not equal the electrons. It is no longer a sodium atom for although its nucleus is unchanged it possesses too few electrons to be a sodium atom. An electrically charged particle such as this is called an **ion** and is written Na^+. Similarly by accepting an electron a **chloride ion** would be formed, and in this case it would carry a negative charge and be written Cl^-.

Once the transference of the electron has taken place, two oppositely charged ions are formed which are capable of attracting each other. An electrostatic force exists between the **ionic pair** resulting in the formation of a very stable crystalline solid, sodium chloride. The process of forming a compound by **transfer of electrons** is called **electrovalence** and the bond formed is called an **electrovalent or ionic bond**.

$Na - e = Na^+$ (sodium ion)
$Cl + e = Cl^-$ (chloride ion)
$Na^+ —— Cl^-$

Each ion now has the electronic structure corresponding to that of a noble gas: the sodium that of neon and the chlorine that of argon.

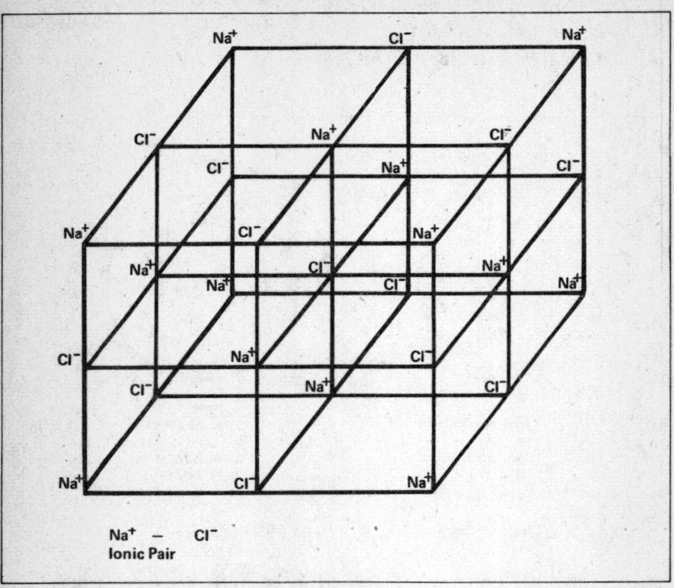

Figure 18. A sodium chloride lattice

Solid sodium chloride consists of sodium ions and chloride ions and an electrostatic force of attraction exists between the ions. A sodium ion is not combined or attached to any particular chloride ion. Each ion of sodium is surrounded by six equidistant chloride ions and vice versa. Figure 18 above illustrates this and indicates the stability of this compound.

Covalent bond formation

Elements may also achieve a stable octet of electrons in their outer shell by **sharing** one or more pairs of electrons. In sharing a pair of electrons one electron is provided by each atom. Each pair of electrons forms a single covalent bond, as, for example, in chlorine (Cl_2).

Each chlorine atom has seven electrons in its outer shell, i.e. one short of the octet. A stable configuration is achieved by each

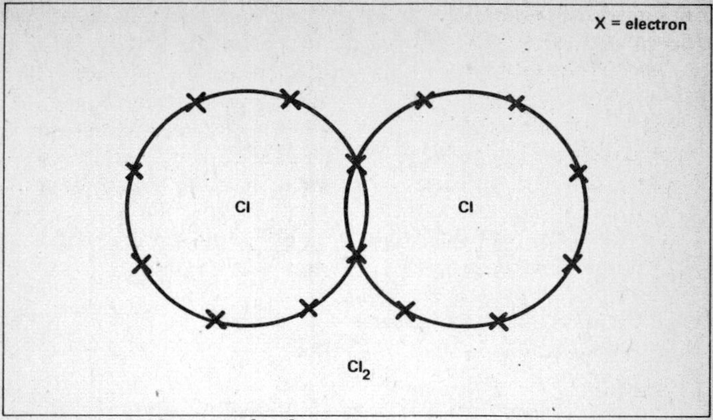

Figure 19. A chlorine molecule

chlorine atom sharing an electron and forming a single covalent bond between the atoms. Hence chlorine is called a **diatomic** gas. Other gases such as hydrogen (H_2), nitrogen (N_2) and oxygen (O_2) form covalent bonds in the same way.

Another good example of a covalent bond is the formation of the compound tetrachloromethane, CCl_4 (carbon tetrachloride)

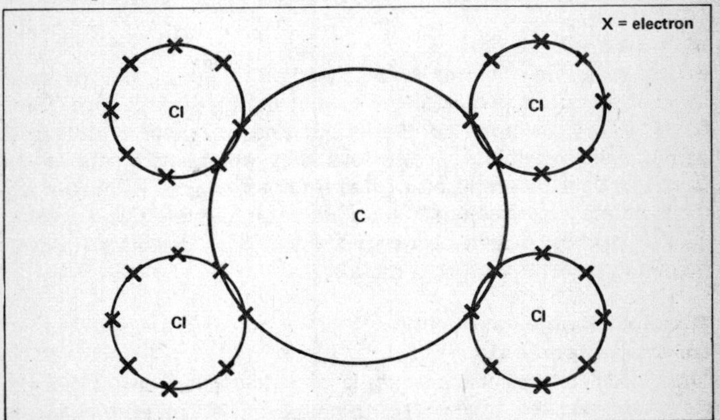

Figure 20. Molecule of tetrachloromethane

Each chlorine atom contributes an electron to be shared with the four electrons possessed by carbon resulting in a stable configuration for each chlorine atom and the carbon atom.

Influence of bond type on the properties of compounds

	Ionic compounds	Covalent compounds
1.	Crystalline solids consisting of a large number of positively and negatively charged ions	Mostly gases or volatile liquids and exist as discrete molecules held together by weak Van der Waals forces.
2.	They possess high melting points and boiling points, owing to the powerful forces of attraction between the charged ions	Usually low melting points and boiling points
3.	Usually very soluble in water forming electrolytes; only slightly soluble in organic solvents	Soluble in organic solvents, but covalent compounds cannot be electrolytes because they do not contain any ions or charged parts
4.	When molten (fused) they are good conductors of electricity because they are composed of ions	Usually non-conductors of electricity

Metal bonding

Metals have high melting points and boiling points, like those of ionic compounds, which suggests that metals should form ionic bonds. They are also good thermal and electrical conductors, which indicates that the electrons they possess are mobile. In fact, metals are composed of metal ions arranged in an orderly lattice through which flows an electron cloud. The ions are closely packed, which accounts for the high densities of most metals compared with non-metals.

Giant molecular structures (macromolecules)

Diamond is an excellent example of a macromolecule. Here the units are bonded together covalently, but form one molecule made up of a large number of atoms.

Such a structure is stable because the atoms are held rigidly together, causing diamond to be extremely hard, with very high melting and boiling points.

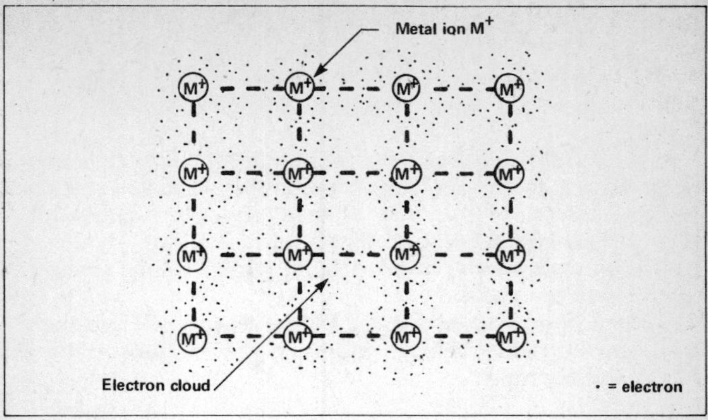

Figure 21. Metal bonding and electron cloud

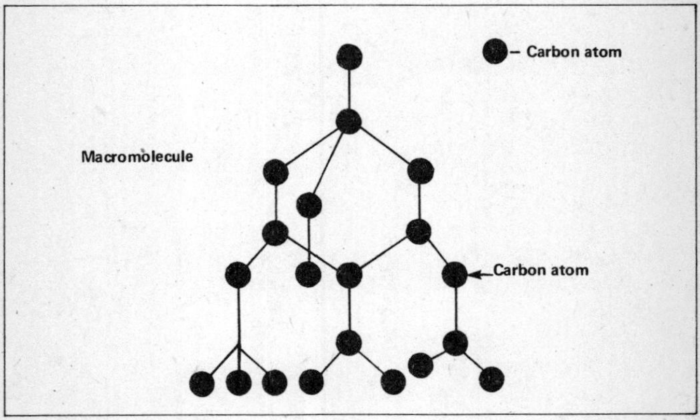

Figure 22. Diamond structure

Key terms

Proton A particle found in the atomic nucleus, it has a charge of +1 and a mass of 1.
Neutron A particle found in the atomic nucleus, it has a charge of zero (neutral) but a mass of 1.
Electron A particle found circulating around the nucleus of the atom. It has a charge of −1 and negligible mass.

Atomic number The charge on the atomic nucleus, i.e. the number of protons in the nucleus of the atom or the number of planetary electrons circulating about that nucleus.
Atomic mass The sum of protons and neutrons in the atomic nucleus.
Atomic weight The average of the atomic masses of the isotopes in an element, taking into consideration their relative abundance.
Isotope Atoms of the same element having the same atomic number but different atomic masses.
Ionic (electrovalent) bond A bond formed by the giving and receiving of electrons.
Covalent bond A bond formed by the sharing of electrons.
Ion An electrically charged atom or radical, formed by the transfer of electrons.

Chapter 6
Oxidation and Reduction

Oxidation can be defined in three ways.

1. By the addition of oxygen to a substance, e.g. magnesium ribbon burns in oxygen to form magnesium oxide.

   $$2Mg(s) + O_2(g) \rightarrow 2MgO(s)$$

   Magnesium has gained or added on oxygen, and in so doing has been **oxidised** to magnesium oxide.
2. By the loss of hydrogen from a substance, e.g. hydrogen sulphide reacts with chlorine to form hydrogen chloride and sulphur.

   $$H_2S(g) + Cl_2(g) \rightarrow 2HCl(g) + S(s)$$

   Hydrogen sulphide has lost hydrogen and in so doing has been **oxidised** to sulphur.
3. By the loss of electrons by a substance, e.g. magnesium reacts with dilute hydrochloric acid to form magnesium chloride and hydrogen.

   $$Mg(s) + 2HCl(aq) \rightarrow MgCl_2(aq) + H_2(g)$$

   This equation can also be written

   $$Mg - 2e \rightarrow Mg^{2+}$$

   Magnesium has lost electrons to become magnesium ions

   OR

   Magnesium has been **oxidised** to magnesium chloride.

Note that the reaction in (1) can also be written in terms of electron transfer.

$$2Mg - 4e \rightarrow 2Mg^{2+}$$
$$2Mg(s) + O_2(g) \rightarrow 2MgO(s)$$

Other examples of **oxidation**

(i) Most elements react with oxygen or air to form the corresponding oxide, e.g.

$$2Zn(s) + O_2(g) \rightarrow 2ZnO(s)$$
$$2Zn - 4e \rightarrow 2Zn^{2+}$$

(ii) Hydrogen burning in oxygen to form water.

$$2H_2(g) + O_2(g) \rightarrow 2H_2O(l)$$
$$2H_2 - 4e \rightarrow 4H^+$$

(iii) Iron(II) compounds being oxidised to iron(III) compounds.

$$Fe^{2+} - e \rightarrow Fe^{3+}$$

(green) (reddish brown)

(iv) In the preparation of chlorine in the laboratory, concentrated hydrochloric acid is oxidised to chlorine. The most common oxidising agent used is manganese(IV) oxide (manganese dioxide), although potassium manganate(VII) (potassium permangate) is more convenient to use because no heat is required. The basic reaction can be conveniently written:

$$2Cl^{-} - 2e \rightarrow Cl_2$$

(v) All anode reactions in electrolysis experiments are oxidation, e.g. the discharge of OH^- ions in the electrolysis of dilute sulphuric acid (page 75).

$$2OH^{-} - 2e \rightarrow H_2O + O$$
$$O + O \rightarrow O_2$$

and copper being converted into copper ions in the electrolysis of copper(II) sulphate using copper electrodes (page 76).

$$Cu - 2e \rightarrow Cu^{2+}$$

Reduction can also be defined in three ways (you will notice that reduction is the absolute reverse of oxidation, i.e. oxidation is to reduction what 'black is to white').

1. By the removal of oxygen from a substance, e.g. copper(II) oxide reacts with hydrogen to form copper and water.

$$CuO(s) + H_2(g) \rightarrow Cu(s) + H_2O(l)$$

i.e. copper oxide has lost oxygen, and in so doing has been reduced to copper.

2. By the gaining of hydrogen by a substance, e.g. nitrogen and hydrogen react together to form ammonia in the Haber process (page 153).

$$N_2(g) + 3H_2(g) \rightleftharpoons 2NH_3(g)$$

i.e. nitrogen has gained hydrogen, and in so doing has been reduced to ammonia.

3. By the gaining of electrons by a substance, i.e. iron oxide is converted into iron in the blast furnace (page 116).

$$Fe_2O_3(s) + CO(g) \rightarrow 2Fe(s) + 3CO_2(g)$$

This can also be written:

$$2Fe^{3+} + 6e \rightarrow 2Fe$$

i.e. iron ions have gained electrons to become iron atoms

OR

iron oxide has been reduced to iron metal.

Note that the reaction in (1) can also be written in terms of electron transfer.

$$Cu^{2+} + 2e \rightarrow Cu$$
$$CuO(s) + H_2(g) \rightarrow Cu(s) + H_2O(1)$$

Other examples of **reduction**

(i) The oxide of zinc, and those of the metals below zinc in the Activity series (page 69) can be converted into metals using reducing agents, e.g.

$$2ZnO(s) + C(s) \rightarrow 2Zn(s) + CO_2(g)$$
$$2Zn^{2+} + 4e \rightarrow 2Zn$$

(ii) Chlorine being reduced to hydrogen chloride by synthesis with hydrogen, e.g.

$$H_2(g) + Cl_2(g) \rightarrow 2HCl(g)$$
$$Cl_2 + 2e \rightarrow 2Cl^-$$

(iii) Iron(III) compounds being reduced to iron(II) compounds.

$$Fe^{3+} + e \rightarrow Fe^{2+}$$

(iv) All cathode reactions in electrolysis experiments are examples of reduction, e.g. in the discharge of copper(II) ions in the electrolysis of copper(II) sulphate solution.

$$Cu^{2+} + 2e \rightarrow Cu$$

Redox (Reduction and Oxidation reactions)

Oxidation and reduction processes always occur together in a consists of $PbBr_2$. It will contain the ions Pb^{2+} and $2Br^-$ (see figure 23).

$$CuO(s) + H_2(g) \rightarrow Cu(s) + H_2O(1)$$

Let us consider what has happened in this reaction to each reactant and product.

1. Copper(II) oxide (CuO) and copper (Cu)

The copper(II) oxide has lost oxygen to become copper. Copper ions have become copper atoms.

$$Cu^{2+} + 2e \rightarrow Cu$$

The copper(II) oxide has been reduced to copper by the hydrogen. Hydrogen must therefore be acting as a reducing agent in this reaction.

2. Hydrogen (H_2) and water (H_2O)

The hydrogen gains oxygen to become water. Hydrogen atoms have become hydrogen ions.

$$H_2 - 2e \rightarrow 2H^+$$

The hydrogen has been oxidised to water by the copper(II) oxide. Copper(II) oxide must therefore be acting as an oxidising agent in this reaction. This reaction can be summarised thus:

$Cu^{2+} + 2e \rightarrow Cu$	(reduction)
$H_2 - 2e \rightarrow 2H^+$	(oxidation)
i.e. $Cu^{2+} + H_2 \rightarrow Cu + 2H^+$	(redox)

In this reaction:

(i) The copper(II) oxide acting as an **oxidising agent** has itself been **reduced** to copper, and

(ii) The hydrogen acting as a **reducing agent** has itself been **oxidised** to water.

Reduction and oxidation have taken place simultaneously in this reaction.

Electrolysis of fused (molten) lead bromide using carbon electrodes

This is an excellent example of a redox reaction. The electrolyte consists of $PbBr_2$. It will contain the ions Pb^{2+} and $2Br^-$ (see figure 23).

To the cathode (−ve) Pb^{2+}	To the anode (+ve) $2Br^-$
The lead ion will be discharged thus:	The bromide ions will be discharged thus:
$Pb^{2+} + 2e \rightarrow Pb$	$2Br^- - 2e \rightarrow Br_2$
Lead will be formed.	Bromine will be formed.

The lead ions have become lead atoms and the bromide ions

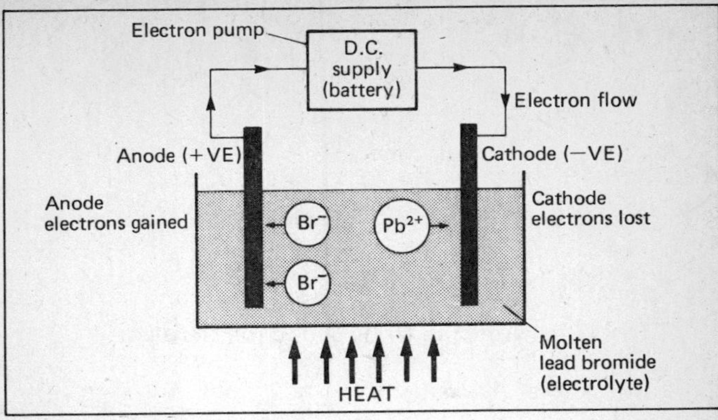

Figure 23. Electrolysis of lead bromide

have become bromine atoms.

$$Pb^{2+} + 2e \rightarrow Pb \quad \text{(reduction)}$$
$$2Br^{-} - 2e \rightarrow Br_2 \quad \text{(oxidation)}$$
$$Pb^{2+} + 2Br^{-} \rightarrow PbBr_2 \quad \text{(redox)}$$

Note that there has been no gain or loss of electrons.

Other examples of redox reactions.

(i) When zinc is added to a solution of copper(II) sulphate, the zinc displaces the copper from the solution.

$$Zn(s) + CuSO_4(aq) \rightarrow ZnSO_4(aq) + Cu(s)$$

This can be written in terms of electron transfer

$$Zn - 2e \rightarrow Zn^{2+} \quad \text{(oxidation)}$$
$$Cu^{2+} + 2e \rightarrow Cu \quad \text{(reduction)}$$
$$Zn + Cu^{2+} \rightarrow Zn^{2+} + Cu \quad \text{(redox)}$$

(ii) When chlorine is bubbled into a solution of sodium bromide, bromine is displaced from the solution.

$$Cl_2(g) + 2NaBr(aq) \rightarrow 2NaCl(aq) + Br_2(1)$$

This can be written in terms of electron transfer.

$$
\begin{array}{llll}
Cl_2 & + 2e & \rightarrow 2Cl^- & \text{(reduction)} \\
2Br^- & - 2e & \rightarrow Br_2 & \text{(oxidation)} \\
\hline
Cl_2 & + 2Br^- & \rightarrow 2Cl^- + Br_2 & \text{(redox)}
\end{array}
$$

(iii) When sodium is burned in chlorine sodium chloride is formed.

$$Na(s) + Cl_2(g) \rightarrow NaCl(s)$$

This can be written in terms of electron transfer.

$$
\begin{array}{ll}
2Na - 2e \rightarrow 2Na^+ & \text{(oxidation)} \\
Cl_2 + 2e \rightarrow 2Cl^- & \text{(reduction)} \\
\hline
2Na + Cl_2 \rightarrow 2NaCl & \text{(redox)}
\end{array}
$$

Key terms

Oxidation is the combination of oxygen with a substance (the gaining of oxygen) or the removal (loss) of hydrogen from it. In terms of electron transfer, oxidation is the loss of electrons by a substance.

Reduction is the removal of oxygen from a substance (the loss of oxygen) or the addition (gain) of hydrogen to it. In terms of electron transfer, reduction is the gaining of electrons by a substance.

Oxidising agent is a substance that gives oxygen to another substance, or removes hydrogen, or removes electrons from it.

Reducing agent is a substance that removes oxygen, or adds hydrogen, or donates electrons to a substance.

Redox reaction This is an oxidation – reduction reaction, in which an oxidising agent has been reduced, and a reducing agent oxidised. This has involved the transfer of electrons from one atom, ion or molecule to another.

Chapter 7
The Electrochemical Series and Electrolysis

Metals differ greatly in the way in which they react, e.g. sodium reacts violently with water and has to be stored in oil to prevent rapid oxidation by the air. Gold does not react with water or air. As a result of the differences of activity between metals it is possible to place the metals in order, putting the most reactive first and least reactive last. Such a list or series is known as the Activity or Electrochemical series (Electromotive series). The series is given below:

Metal	Metal
1. Potassium	8. Tin
2. Sodium	9. Lead
3. Calcium	10. (Hydrogen) – non-metal
4. Magnesium	11. Copper
5. Aluminium	12. Silver
6. Zinc	13. Gold
7. Iron	

The above list can be justified by studying the following:
1. the action of oxygen (air) on the metals;
2. the action of water on the metals;
3. the action of acids on the metals;
4. the displacement of metals by other metals in the series;
5. the compounds of the metals.

1. Action with oxygen (air)

Potassium and sodium tarnish immediately when exposed to the air, each forming oxides.
Calcium and magnesium will burn readily in air or oxygen while **aluminium, zinc and iron** have to be heated very strongly to burn in oxygen.
Tin, lead and copper all oxidise when strongly heated but do not burn.
Silver and gold form oxides with very great difficulty.

2. Action with water

Potassium reacts violently with **cold water**, burning with a characteristic lilac flame. Hydrogen is displaced from the water and the very strong alkali (potassium hydroxide) is formed. During the course of the reaction the potassium darts around in a haphazard way on the surface of the water.

$$2K(s) + 2H_2O(l) = 2KOH(aq) + H_2(g)\uparrow$$

Sodium reacts almost as violently with **cold water** as potassium burning with a yellow flame, hydrogen and sodium hydroxide being the products of the reaction.

$$2Na(s) + H_2O(l) = 2NaOH(aq) + H_2(g)\uparrow$$

Calcium reacts readily with **cold water**, but not as violently as potassium or sodium. The calcium sinks to the bottom of the container holding the water, hydrogen gas being evolved at a steady rate, calcium hydroxide solution (limewater) being the other reaction product.

$$Ca(s) + 2H_2O(l) = Ca(OH)_2(aq) + H_2(g)\uparrow$$

Note Potassium, sodium and calcium react with cold water to form alkalis (soluble bases).

Magnesium reacts very slowly with hot water, but will react much quicker when heated in **steam**.

$$Mg(s) + H_2O(g) = MgO(s) + H_2(g)\uparrow$$

Iron at red heat will react with steam to form the black oxide of iron and hydrogen. This reaction is reversible.

$$3Fe(s) + 4H_2O(g) \rightleftharpoons Fe_3O_4(s) + 4H_2(g)\uparrow$$

Note Magnesium, aluminium, zinc and iron react with hot water or steam to form basic oxides. Water and steam have no action with the metals below iron (no. 7) in the series.

3. Action with acids

Potassium, sodium and calcium will displace hydrogen from dilute hydrochloric and sulphuric acids with explosive violence. All three would produce very dangerous reactions.

Zinc, iron and magnesium (Z.I.M. metals) will displace hydrogen from dilute hydrochloric and sulphuric acids. Moderate reactions occur.

$$Mg(s) + 2HCl(aq) = MgCl_2(aq) + H_2(g)\uparrow$$
$$Zn(s) + H_2SO_4(aq) = ZnSO_4(aq) + H_2(g)\uparrow$$
$$Fe(s) + 2HCl(aq) = FeCl_2(aq) + H_2(g)\uparrow$$

The reaction between zinc and dilute sulphuric acid is commonly used in the laboratory preparation of hydrogen.

Lead will only displace hydrogen from hot concentrated hydrochloric acid.

Copper, silver and gold being lower (less electropositive) than hydrogen in the series will not displace it from acids.

4. Displacement of metals by other metals

All metals above hydrogen in the series will displace it from some acids. The hydrogen in sulphuric and hydrochloric acids is displaced by metals which are more electropositive than hydrogen or more active and therefore higher placed in the series. Metals below hydrogen in the series do not displace it from acids. A general rule can be derived from these facts: namely, that one metal will displace another metal lower in the series from solutions of its salts, whilst the higher-placed metal would itself be displaced by metals above it in the series, i.e. the more electropositive metal will displace the less electropositive metal from its salt. Moreover the greater the gap between metals in the series, the easier will be the displacement.
E.g. Magnesium would displace copper from a solution of copper (II) sulphate far more easily than lead. On the other hand it would be far easier for magnesium to displace copper from copper (II) sulphate than to displace zinc from zinc sulphate.

Examples

$$Zn + 2HCl = ZnCl_2 + H_2\uparrow$$
$$Zn + 2H^+ = Zn^{2+} + H_2\uparrow$$

$$Fe + CuSO_4 = FeSO_4 + Cu\downarrow$$
$$Fe + Cu^{2+} = Fe^{2+} + Cu\downarrow$$

$$Zn + CuSO_4 = ZnSO_4 + Cu\downarrow$$
$$Zn + Cu^{2+} = Zn^{2+} + Cu\downarrow$$

$Cu + ZnSO_4$ — no reaction

5. Compounds of metals

The more electropositive a metal, the more stable compounds it tends to form, e.g. one of the most stable compounds is sodium chloride (common salt).
Examination of the compounds of some of the metals justifies the above statement.

Oxides

The lower down they appear in the series, the more easily can the oxides of the metals be reduced, e.g. the oxides of the metals down to, and including, aluminium oxide cannot be reduced to metals except by electrical means.

Zinc is the first metal of which the oxide can be reduced by means other than electrolysis.

$$2ZnO + C = 2Zn + CO_2$$

Iron is extracted in blast furnaces by means of reducing its ore (oxide) with the aid of carbon monoxide.

$$Fe_2O_3 + 3CO = 2Fe + 3CO_2$$

Copper (II) oxide can be reduced in the laboratory by heating in a stream of hydrogen gas.

$$CuO + H_2 = Cu + H_2O$$

Nitrates

All metal nitrates break down under varying degrees of heat to produce oxygen. The nitrates of potassium and sodium decompose into the respective nitrites and oxygen.

$$2KNO_3 = 2KNO_2 + O_2\uparrow$$

Most of the other nitrates decompose on heating to form the respective oxides, nitrogen dioxide and oxygen.

$$2Pb(NO_3)_2 = 2PbO + 4NO_2\uparrow + O_2\uparrow$$
$$2Cu(NO_3)_2 = 2CuO + 4NO_2\uparrow + O_2\uparrow$$

Carbonates

The most stable carbonates are those of potassium and sodium, both of which remain unchanged when heated. Calcium carbonate will break down under strong heat to form calcium oxide (quicklime) and carbon dioxide.

$$CaCO_3 = CaO + CO_2\uparrow$$

Copper (II) carbonate will break down very easily upon heating into copper (II) oxide and carbon dioxide.

$$CuCO_3 = CuO + CO_2\uparrow$$

Electrolysis

Electrolysis is the decomposition of a compound (molten or in solution) by means of passing an electric current through it.

Compounds which undergo **electrolysis** either fused (molten) or in solution are called **electrolytes.** Compounds which do not undergo electrolysis are called **non-electrolytes**.

All electrolytes contain ions, which are electrically charged particles formed from an atom or a group of atoms by the gain or loss of electrons.

$Cu - 2e$	$=$	Cu^{2+}
copper atom		copper ion

Some students have difficulty in expressing the formulae of the ions present in a particular electrolyte. It is useful in this context to remember that metals form positive ions (cations) whereas non-metals usually form negative ions (anions), and also that the number of charges carried by an ion is the same as the valency of the atom or group from which it is derived.

Hydrogen usually forms positive ions. Some examples of the ions present in common electrolytes are shown below.

Sodium chloride (NaCl)	—	Na^+	$+ Cl^-$
Sodium sulphate (Na_2SO_4)	—	$2Na^+$	$+ SO_4^{2-}$
Copper (II) sulphate ($CuSO_4$)	—	Cu^{2+}	$+ SO_4^{2-}$
Lead(II) bromide ($PbBr_2$)	—	Pb^{2+}	$+ 2Br^-$
Sulphuric acid (H_2SO_4)	—	$2H^+$	$+ SO_4^{2-}$

In an electrolyte the ions are present before the electric current is passed.

E.g. A solution of sodium chloride would contain Na^+, Cl^-, H^+ and OH^- ions. If this particular solution is electrolysed, the positive ions (cations) would migrate to the negative electrode (cathode) while the negative ions (anions) would migrate toward the positive electrode (anode).

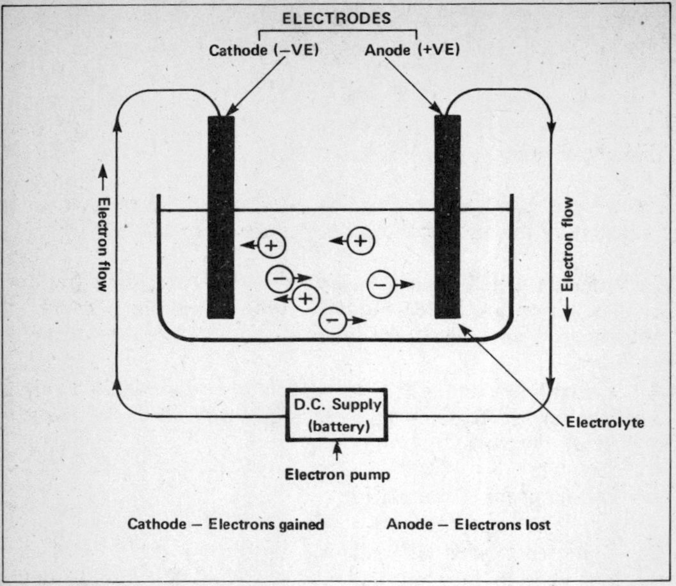

Figure 24. Electrolysis cell

Once arriving at the electrodes the ions are discharged, i.e. they become neutral atoms again.

E.g.	$Cu^{2+} + 2e$	$=$	Cu
	Copper ion		Copper atom

Certain rules govern which ion will be discharged first if more than one kind of ion is present (see below).

Factors affecting the discharge of ions

1. The position of the ion in the series

The ion which is lowest in the series is discharged first.

Cations	Anions
K^+	SO_4^{2-}
Na^+	NO_3^-
Ca^{2+}	Cl^-
|	Br^-
↓	I^-
Au^+	OH^-

If this is difficult to understand, think what happens when a cation is discharged. Electrons are accepted by the cation.
E.g. $Cu^{2+} + 2e = Cu$

The most reactive metals at the top of the series form ions most readily because the electronic configuration of the ion is more stable than that of the atom. Hence the ion is reluctant to accept electrons and reform an atom, and therefore remains in solution in preference to the discharge of an ion lower in the series.

2. The concentration of ions
If one ion has a much greater concentration than the others, despite its position in the series, it may be preferentially discharged.
E.g. Electrolysis of concentrated sodium chloride solution yields chlorine at the anode, whereas dilute sodium chloride solution yields oxygen.
I.e. OH^- ions are only preferentially discharged when Cl^- ions are at a low concentration.

3. The type of electrodes
Inert electrodes, e.g. platinum and carbon, do not affect the choice of ion for discharge, but other electrodes may do so.
E.g. Electrolysis of copper sulphate using copper and platinum electrodes.

Examples of electrolysis

1. Electrolysis of dilute sulphuric acid by means of platinum or carbon electrodes
The electrolyte consists of H_2SO_4 and H_2O. It will therefore contain the ions:

$2H^+$ SO_4^{2-} (H_2SO_4)	H^+ OH^- (H_2O)
To the anode:	To the cathode:
SO_4^{2-}, OH^-	H^+
ions discharged OH^- anode reaction	ions discharged H^+ cathode reaction
$OH^- = OH + e$	$H^+ + e = H$

$2OH^- = H_2O + O$ $H + H = H_2\uparrow$

$O + O = O_2\uparrow$

Oxygen gas liberated — Hydrogen gas liberated

Summary

H^+ ions are discharged at the cathode forming hydrogen gas. At the anode OH^- ions are discharged in preference to SO_4^{2-} ions, requiring less energy to do so, and resulting in the formation of oxygen gas.

2. Electrolysis of copper (II) sulphate solution

(a) **By means of platinum electrodes**

The electrolyte consists of $CuSO_4$ and H_2O. It will contain the ions

Cu^{2+} SO_4^{2-} H^+ OH^-

$(CuSO_4)$ (H_2O)

To the anode SO_4^{2-}, OH^- ions discharged OH^- anode reaction

$OH^- \rightarrow OH + e$

$2OH \rightarrow H_2O + O$

$O + O \rightarrow O_2\uparrow$

Oxygen gas liberated

To the cathode Cu^{2+}, H^+ ions discharged Cu^{2+} cathode reaction

$Cu^{2+} + 2e \rightarrow Cu$

Copper metal deposited

Summary

Cu^{2+} ions are preferentially discharged at the cathode forming a coating of copper metal. OH^- ions are discharged at the anode forming oxygen gas.

(b) **By means of copper electrodes**

The electrolyte contains the same ions as in (a).

To the anode SO_4^{2-}, OH^- ions discharged Cu^{2+} (from the active electrode)
anode reaction

$Cu \rightarrow Cu^{2+} + 2e$

(requiring least amount of energy)

To the cathode Cu^{2+}, H^+ ions discharged Cu^{2+}

cathode reaction

$Cu^{2+} + 2e = Cu$

Copper deposited

Summary

Cu^{2+} ions are preferentially discharged at the cathode forming a coating of copper metal. At the anode copper atoms from the anode itself give up electrons forming Cu^{2+} since this process requires less energy than the discharge of OH^- ions.

3. Electrolysis of sodium sulphate solution by means of carbon electrodes

The electrolyte consists of Na_2SO_4 and H_2O. It will contain the ions:

$2Na^+$, SO_4^{2-} — (Na_2SO_4)

H^+, OH^-, — (H_2O)

To the anode SO_4^{2-}, OH^- ions
discharged OH^-
anode reaction

$$OH^- \rightarrow OH + e$$

$$2OH \rightarrow H_2O + O$$

$$O + O \rightarrow O_2\uparrow$$

Oxygen gas liberated

To the cathode $2Na^+$, H^+ ions
discharged H^+
cathode reaction

$$H^+ + e \rightarrow H$$

$$2H \rightarrow H_2\uparrow$$

Hydrogen gas liberated

Summary

H^+ ions are discharged at the cathode forming hydrogen gas. At the anode OH^- ions are preferentially discharged to SO_4^{2-} ions, requiring less energy to do so, resulting in the formation of oxygen gas.

Faraday's laws of electrolysis

Michael Faraday's first law states that the weight of a substance produced during electrolysis is directly proportional to the quantity of electricity passed through the electrolyte. When the quantity of electricity used is 1 Faraday (96,500 coulombs, which is 26.8 ampère hours) then a mole of monovalent ions would be liberated, or half a mole of divalent ions or a third of a mole of trivalent ions. It therefore follows that the masses of different substances liberated by the same quantity of electricity will be proportional to their atomic weights or to some simple fraction of the atomic weight depending on the valency of the ion discharged.

Example
What weight of silver would be liberated by the same quantity of electricity which liberates 0.32g of copper from copper (II) sulphate solution?

Answer Copper is divalent while silver is monovalent. Thus 1 Faraday (96,500 coulombs) of electricity would deposit 1 mole of silver ions, i.e. 108g (atomic weight) of silver. Since copper is divalent, 1 Faraday of electricity would deposit $\frac{1}{2}$ mole of copper, i.e. 64/2g = 32g (64 is the atomic weight of copper). Thus for every 32g of copper, 108g of silver are deposited. ∴ if 0.32g of copper is deposited, 1.08g of silver must be formed.

The Faraday
The Faraday is a mole of electrons, i.e. the quantity of electricity which would be carried by 6×10^{23} (Avogadro number) electrons.

Key terms

Activity (Electrochemical or Electromotive) series A table of metallic elements placed in order of their reactivity.
Anode The positive electrode.
Cathode The negative electrode.
Electrode The conductor by which the electric current passes into or out of a solution during electrolysis.
Electrolyte A solution or fused salt which conducts the electric current and is decomposed by it.
Faraday A mole of electrons, i.e. the quantity of electricity which would be carried by the Avogadro number of electrons (6×10^{23}).

Chapter 8
Acids, Bases and Salts

Acids

Non-metals burn in air (oxygen) to form **oxides**, e.g. the element sulphur burns in air to form the gas sulphur dioxide. This gas will dissolve in water to form the weak sulphurous acid.

$$S(s) + O_2(g) = SO_2(g)\uparrow$$
$$SO_2(g) + H_2O(l) = H_2SO_3(aq)$$

Acids can be solids, liquids or gases, e.g. citric acid which is found in the juice of oranges, lemons and limes, is in the pure form a solid. Sulphuric acid is made by bubbling sulphur trioxide into strong sulphuric acid producing concentrated sulphuric acid, which is a heavy oily liquid. Hydrochloric acid is a gas dissolved in water.

The three acids most commonly used in schools are hydrochloric, sulphuric and nitric acids, these being referred to as the **mineral acids**.

Properties of acids

1. They all have a sour taste.
2. In solution they turn blue litmus red.
3. They contain hydrogen, some or all of which can be replaced by a metal to form a **salt.**
   E.g. Sulphuric acid has two replaceable hydrogens – H_2SO_4, or is said to have a **basicity** of 2. If one of these hydrogen atoms is replaced by a sodium atom, then sodium hydrogen-sulphate (sodium bisulphate) is formed, which is called the **acid salt.**

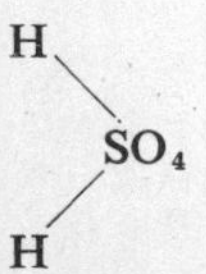

acid

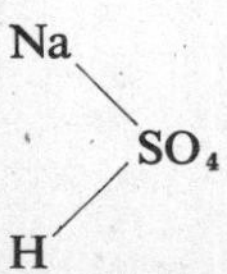

acid salt
sodium hydrogensulphate

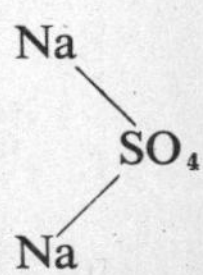

normal salt
sodium sulphate

If both hydrogens are displaced sodium sulphate, the **normal salt**, is formed. These salts are prepared by the action of sodium hydroxide solution on dilute sulphuric acid.

$$NaOH(aq) + H_2SO_4(aq) = NaHSO_4(aq) + H_2O(1)$$
$$2NaOH(aq) + H_2SO_4(aq) = Na_2SO_4(aq) + 2H_2O(1)$$

4. They liberate carbon dioxide from carbonates and hydrogen-carbonates.

$$Na_2CO_3(s) + 2HCl(aq) = 2NaCl(aq) + H_2O(1) + CO_2(g)\uparrow$$

carbonate + acid = salt + water + carbon dioxide

5. They will react with some metals, e.g. **z**inc, **i**ron, **m**agnesium (**Z.I.M.** metals) to form salts with the liberation of hydrogen gas.

$$Zn(s) + H_2SO_4(aq) = ZnSO_4(aq) + H_2(g)\uparrow$$
$$Zn + 2H^+ = Zn^{2+} + H_2\uparrow$$

It is worth noting that when metals such as sodium and potassium displace hydrogen from acids, they do so with dangerous violence. Nitric acid reacts with most metals but will only produce hydrogen with magnesium, and only then when the acid is very, very dilute.

6. They can be neutralised by basic oxides and **alkalis** (soluble basic oxides) to form salts and water.

$$CuO + H_2SO_4 = CuSO_4 + H_2O$$

Basic oxide + acid = salt + water

7. They all yield hydrogen ions when dissolved in water, i.e. they all **ionise** to a greater or lesser degree.

E.g. $HCl \rightleftharpoons H^+ + Cl^-$

$H_2SO_4 \rightleftharpoons 2H^+ + SO_4^{2-}$

Strong acids such as hydrochloric and sulphuric ionise almost completely, while weak acids such as the organic acids are only partially ionised, e.g. acetic acid (CH_3COOH), the acid found in vinegar, ionises thus:

$CH_3COOH \rightleftharpoons CH_3COO^- + H^+$

but does so only partially.

When an acid ionises the hydrogen ions liberated are hydrated and are known as **hydroxonium ions,** H_3O^+. When hydrochloric acid ionises it does so thus:

$$HCl + H_2O \rightleftharpoons H_3O^+ + Cl^-$$

An acid could therefore be defined as a substance which when dissolved in water forms hydroxonium ions (H_3O^+) as the only positively charged ions.

Acids could by this definition be regarded as **proton donors** in that their tendency is to 'donate an H^+'. Acids found in everyday life include:

carbonic acid in soft drinks;
formic acid (HCOOH) in ants and nettles;
acetic acid (CH_3COOH) in vinegar;
citric acid ($C_6H_8O_7$) in citrus fruits;
tartaric acid $(CHOHCOOH)_2$ in baking powder.

Preparation of acids

1. By dissolving the oxide of a non-metal in water:
   E.g. $SO_2(g) + H_2O(l) = H_2SO_3(aq)$ (sulphur**ous** acid)
   $SO_3(g) + H_2O(l) = H_2SO_4(aq)$ (sulphur**ic** acid)

2. By heating the salt of a volatile acid with a non-volatile acid:
   E.g. Hydrogen chloride can be prepared by heating common salt (sodium chloride) with concentrated sulphuric acid.
   $NaCl + H_2SO_4 = NaHSO_4 + HCl$
   Hydrogen chloride is commonly called **salt gas** because it is prepared from common salt.
   Nitric acid is produced by heating a suitable nitrate with concentrated sulphuric acid.
   $2NaNO_3 + H_2SO_4 = Na_2SO_4 + 2HNO_3$

Bases and alkalis

A base is a substance which will react with an acid to form a salt and water, and one which contains **oxide** (O^{2-}) or **hydroxyl** (OH^-) ions.

An alkali is a soluble base, and forms hydroxyl ions (OH^-) as the only negatively charged ions.

When a base reacts with an acid it forms a salt, and in doing so it must accept the proton donated by the acid.

E.g. Consider the reaction between solutions of hydrochloric acid and sodium hydroxide.

$$NaOH + HCl = NaCl + H_2O$$

This reaction could also be written:

$$H_3O^+ + Cl^- + Na^+ + OH^- = 2H_2O + Cl^- + Na^+$$

As both sides of the equation contain Na^+ and Cl^- the reaction can be summarised as:

$H_3O^+ + OH^- = 2H_2O$
acid + base
proton donor + proton acceptor

Acids are **neutralised** by bases or alkalis:

E.g. CuO	$+ H_2SO_4$	$= CuSO_4$	$+ H_2O$
basic oxide	+ acid	= salt	+ water
NaOH	+ HCl	= NaCl	$+ H_2O$
alkali	+ acid	= salt	+ water

Some common alkalis are:
sodium hydroxide (NaOH) – caustic soda;
potassium hydroxide (KOH) – caustic potash;
calcium hydroxide $Ca(OH)_2$ – in solution called limewater;
ammonium hydroxide (NH_4OH) – 'ammonia'.

The pH scale

The strengths of the various acids or bases can be measured according to the **pH scale.** Pure water is ionised only slightly

$$H_2O \rightleftharpoons H^+ + OH^-$$

and 1 litre (dm^3) of water contains 10^{-7} moles of hydrogen ions and 10^{-7} moles of hydroxyl ions. It is inconvenient to measure hydrogen ion concentration in mole/dm^3, therefore a logarithmic scale is used resulting in numbers from 1 to 14. A pH value of 7 would indicate that a solution is neutral, a figure less than 7 acid, whilst one more than 7 would indicate that the solution is alkaline.

E.g. Sulphuric acid being a very strong acid would have a pH of 1; caustic soda solution being a strong alkali would have a pH of 14. Weaker acids and alkalis would have respective values between 1 and 14.

To find the pH of a solution or liquid quickly a mixture of

indicators known as Universal Indicator is used, the pH of the solution corresponding to a particular colour (see figure 25).

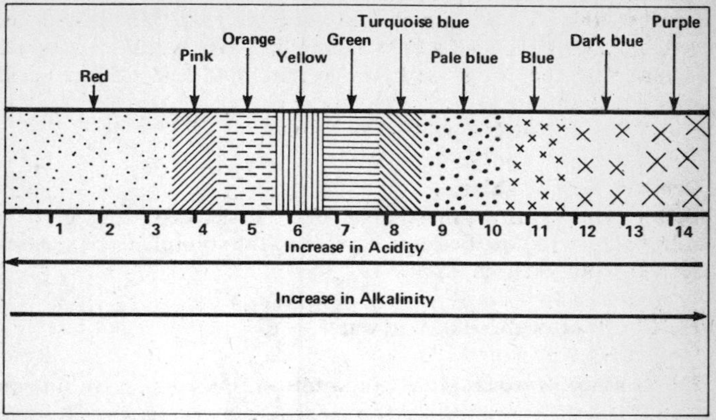

Figure 25. Colours of Universal Indicator

Use is made of Universal Indicator papers by doctors in testing urine and other samples for acidity or alkalinity. Soil-test kits also include an indicator whereby the pH of the soil can be ascertained. A knowledge of the alkalinity or acidity of soils is important in agriculture.

Indicators

These substances are usually vegetable dyes that take on different colours in acid and alkali solutions. Common indicators include red and blue litmus, phenolphthalein, methyl orange and screened methyl orange. Below is a table of the colours shown by various indicators in acid and alkali solutions.

Indicator	Colour in acids	Colour in alkalis
Red litmus	red	blue
Blue litmus	red	blue
Phenolphthalein	colourless or white	pink
Methyl orange	red	yellow
Screened methyl orange	red	green

Salts

When the hydrogen atoms in an acid are replaced by metal atoms (or the NH_4 radical) a new substance with different properties is formed, and is called a **salt.** Common **salt** or sodium chloride is made by replacing the hydrogen atom in hydrochloric acid with a sodium atom.

HCl **Na**Cl

Preparation of salts

Before preparing a salt it must first be decided whether it is soluble or insoluble. Below is a guide to the solubilities of various salts and other compounds.

1. All **nitrates** are soluble in water.

2. All **sodium potassium** and **ammonium** salts are soluble in water.

3. All **hydrogencarbonates** (bicarbonates) are soluble in water (remember that the only solid hydrogencarbonates are those of sodium and potassium).

4. All **chlorides** are soluble in water except lead and silver chlorides.

5. All **sulphates** are soluble in water except those of lead and barium. Calcium sulphate is only sparingly soluble in water.

6. All **carbonates** are insoluble in water except those of sodium, potassium and ammonium.

7. All **metal oxides** are insoluble in water except sodium oxide (Na_2O), potassium oxide (K_2O) and calcium oxide (CaO), which react to form the alkalis sodium hydroxide ($NaOH$), potassium hydroxide (KOH) and calcium hydroxide ($Ca(OH)_2$).

8. All **metal hydroxides** are insoluble except sodium hydroxide, potassium hydroxide and ammonium hydroxide (ammonia gas dissolved in water). Calcium hydroxide (solid) is sparingly soluble in water and forms calcium hydroxide solution (limewater).

The main methods of preparing salts are the following.

1. Making **insoluble** salts by **precipitation** involving **double decomposition.** This method is used for most carbonates, sulphides and lead salts, together with barium sulphate and silver chloride, and can also be used to prepare most hydroxides.

2. Making soluble salts of metals other than sodium, potassium and ammonium. Typical examples include the preparation of:
(a) copper (II) sulphate from copper (II) oxide;
(b) iron (II) sulphate from iron filings;
(c) lead nitrate from lead carbonate;
When the chemical reaction has taken place, the salt is obtained by filtration and crystallisation.

3. Making salts of sodium, potassium and ammonium from their hydroxides and carbonates. This method involves using burettes and pipettes with the indicator methyl orange, screened methyl orange or litmus.

4. Making acid salts, in particular the hydrogensulphates (bisulphates) of sodium and potassium, and the hydrogencarbonates (bicarbonates) of sodium, potassium and calcium.

Method 1
The preparation of an insoluble salt
To make the insoluble salt XY, two solutions are required: one, say XA, containing the cation of X and the other, say BY, containing the anion of Y (where XA and BY are both soluble). Then $XA + BY = XY\downarrow + AB$.
E.g. In the preparation of the insoluble salt lead sulphate the solutions used are lead nitrate (soluble salt of lead) and a soluble sulphate (sodium, potassium, etc.).

Lead nitrate + sodium sulphate = lead sulphate ↓ + sodium nitrate

$$(XA + BY = XY\downarrow + AB)$$
$$Pb(NO_3)_2 + Na_2SO_4 = PbSO_4\downarrow + 2NaNO_3$$

Experimental details
Add the sodium sulphate solution to the lead nitrate solution. A white precipitate of lead sulphate will form. Allow this

precipitate to settle and carefully add to the **supernatant liquid** (the liquid on top of the precipitate) a few more drops of solution to ensure that all the lead has been precipitated. Filter or centrifuge, and wash the precipitate with water to free it from contaminating liquids. Dry the solid in an oven, or between filter papers if it decomposes on heating.

When making carbonates use aqueous sodium carbonate solution to provide the anion; for sulphides, bubble in hydrogen sulphide gas; for hydroxides use aqueous ammonium hydroxide.

Examples:

$ZnSO_4 + Na_2CO_3 = Na_2SO_4 + ZnCO_3\downarrow$

$Pb(NO_3)_2 + H_2S = PbS\downarrow + 2HNO_3$

$FeCl_3 + 3NH_4OH = Fe(OH)_3\downarrow + 3NH_4Cl$

Method 2

The preparation of soluble salts of metals (other than those of sodium or potassium)

Warm the appropriate dilute acid in a porcelain (crystallising) dish, add small portions of the metal, metal oxide, or metal carbonate until the solid is in excess. There is an important reason for this: if there is excess solid at the end of the reaction it is a simple matter to separate the solid from the salt solution; however, if there is excess acid it is far more difficult to separate the solution of the salt from the acid. Filter and evaporate the filtrate until a small portion yields crystals on cooling. Allow the main portion of the solution to stand for a few hours so that crystals form. Associated with the crystals is a liquid called the **mother liquor.** Pour this away, and dry the crystals between filter papers.

Examples:

$Fe + H_2SO_4 = FeSO_4 + H_2\uparrow$ (crystals obtained are $FeSO_4 7H_2O$)

$PbO + 2HNO_3 = Pb(NO_3)_2 + H_2O$ (crystals are $Pb(NO_3)_2$)

$Ca(OH)_2 + 2HNO_3 = Ca(NO_3)_2 + 2H_2O$

$CuCO_3 + 2HCl = CuCl_2 + CO_2\uparrow + H_2O$

Method 3

The preparation of salts of sodium potassium and ammonium from their hydroxides

Make a solution of the appropriate hydroxide or carbonate. Measure 25 cm^3 with a pipette and colour the solution with a few drops of methyl orange or another appropriate indicator. Add from the burette the appropriate dilute acid until the colour of the indicator just changes, i.e. the addition of one drop of acid changes the colour. When this stage is reached the acid has neutralised the alkali (or carbonate) to form a salt.

Acid + alkali = salt + water

The resultant solution if left to crystallise out would yield crystals which would be stained by the indicator. The experiment should be repeated without any indicator (the volume of acid added being determined from the burette reading). Crystals are then obtained as in Method 2.

Examples:

$$KOH + HNO_3 = KNO_3 + H_2O$$
alkali + acid = salt + water

$$K_2CO_3 + 2HCl = 2KCl + CO_2 + H_2O$$
carbonate + acid = salt + carbon dioxide + water

Method 4

The preparation of an acid salt

(i) Making potassium hydrogensulphate (bisulphate). Use Method 3 but at the stage where no indicator is added, add twice the volume of acid needed to turn the indicator or half of the volume of alkali.

$$2KOH + H_2SO_4 = K_2SO_4 + 2H_2O$$
$$KOH + H_2SO_4 = KHSO_4 + H_2O$$

or

$$\mathbf{2}KOH + H_2SO_4 = K_2SO_4 + 2H_2O$$
$$\mathbf{2}KOH + \mathbf{2}H_2SO_4 = \mathbf{2}KHSO_4 + \mathbf{2}H_2O$$

The first pair of equations shows the ratio of alkali to be 2:1, i.e. to make the acid salt half the volume of alkali would be used compared with that for the preparation of the normal salt. The second pair of equations shows the acid ratio to be 1:2, hence the acid salt can also be made by adding twice the volume of acid compared with the volume required for the normal salt.

The addition of twice the volume of acid is the most convenient in that this is usually the variable quantity, i.e. 25 cm^3 of alkali is used as a standard aliquot.

(ii) Making hydrogencarbonates using the acid anhydride. This method is really the same as 4(i) in that the acid salt used is excess carbon dioxide, which takes the place of the acid. Place the alkali in a flask and bubble carbon dioxide into it for a comparatively long time (3–4 minutes). If the alkali is **boiling** the **normal salt** will be formed, but if the alkali is **cold** the **acid salt** will be formed if such a salt exists.

Examples:

Boiling $CO_2 + 2NaOH = Na_2CO_3 + H_2O$
Cold $CO_2 + NaOH = NaHCO_3$
Cold $CO_2 + H_2O + Na_2CO_3 = 2NaHCO_3$
Boiling $CO_2 + Ca(OH)_2 = CaCO_3 + H_2O$
Cold $2CO_2 + Ca(OH)_2 = Ca(HCO_3)_2$

Key terms

Oxide The type of compound formed between an element and oxygen, e.g. when sulphur combines with oxygen, sulphur dioxide (SO_2) or sulphur trioxide (SO_3) is formed. The rust which forms on iron is iron oxide (Fe_2O_3).

Acidic oxides These are the **oxides of non-metals** which dissolve in water to form an **acid** solution.

E.g. $SO_2 + H_2O = H_2SO_3$
$CO_2 + H_2O = H_2CO_3$

Basic oxides These are the **oxides of metals.** Those that dissolve in water, i.e. soluble bases, are known as **alkalis.** Thus if a basic oxide dissolves in water, the solution is alkaline.

Amphoteric oxides These oxides behave as acidic and basic oxides, e.g. Al_2O_3, ZnO, PbO.

Neutral oxides These are the oxides of non-metals which react neither with acids nor with alkalis. They include carbon monoxide (CO), nitrogen monoxide (NO) and water.

Peroxides Peroxides liberate hydrogen peroxide when treated with dilute acids.

E.g. $BaO_2 + H_2SO_4 = H_2O_2 + BaSO_4$

Acid A compound which contains hydrogen ions as the only positive ions.

E.g. $H_2SO_4 \rightleftharpoons 2H^+ + SO_4^{2-}$

Acid salt A compound in which part but not all of the replaceable hydrogen atoms in a molecule of an acid have been replaced by a metal or the ammonium radical, e.g. Na**H**SO_4, Na**H**CO_3.

Normal salt Salt formed when all the replaceable atoms of hydrogen in a molecule of an acid have been replaced by a metal or the ammonium radical, e.g. Na_2SO_4, Na_2CO_3.

Acid anhydride An acidic oxide which will react with water to form an acid solution, e.g. sulphur trioxide (SO_3) is the acid anhydride of sulphuric acid H_2O, SO_3 (H_2SO_4).

$$SO_3 + H_2O = H_2SO_4$$

Basicity The number of replaceable hydrogen atoms contained in one molecule of the acid.

E.g. H_2SO_4 – basicity of 2; H_3PO_4 – basicity of 3;
CH_3COOH – basicity of 1; HCl – basicity of 1.

Hydroxonium ion When an acid ionises, the hydrogen ions liberated are known as hydroxonium ions H_3O^+.

E.g. $HCl + H_2O \rightleftharpoons H_3O^+ + Cl^-$

Base A substance which will react with an acid to form only a salt and water, and one which contains oxide (O^{2-}) or **hydroxyl** (OH^-) ions.

Alkali A soluble base which forms hydroxyl ions (OH^-) as the only negatively charged ions.

Indicators Chemical compounds, usually vegetable dyes, which by changing their colour will indicate whether a solution is acidic or alkaline. The most common and best-known indicator is probably litmus.

Supernatant liquid The liquid found above a precipitate after the precipitate has settled.

Chapter 9
The Periodic Classification of the Elements

Whether one is attempting to classify elements, species or any other entities, the aim must always be to group together those which resemble each other, and separate those which are different from one another.

Dobereiner and Newlands were among the first scientists to attempt to classify the elements and to place them in some sort of logical order. However, the most successful of the earlier classifications was made by Dmitri Mendeleeff, a Russian scientist who in 1869 placed the then known elements in order of their atomic weights. His effort was a major breakthrough in this particular field and indeed the modern version of the Periodic table has for its basis Mendeleeff's original.

The Russian noticed that the properties of the elements generally repeated themselves at regular intervals. He arranged elements with similar properties vertically beneath each other to form groups. More than once when arranging the elements he found that there were some which were obviously misplaced. Realising that some of the misplacements were caused by missing (undiscovered) elements, he therefore, with great wisdom, left gaps in his table, so that once his work was complete and the overall pattern had emerged, he was able to forecast the properties of the undiscovered elements with great accuracy simply through his knowledge of the other elements in the same 'family' or group.

He stated that **the properties of elements are a periodic function of their atomic weights,** and thus gave to science the first statement of the Periodic law.

Nowadays the Periodic table is determined by atomic number and not atomic weight. There is a fundamental quantity which increases in regular steps from one element to the next in the Periodic table. This quantity can only be the charge on the nucleus, i.e. the atomic number or the number of electrons in the cloud circulating around the nucleus. It is the charge on the nucleus alone which determines the chemical properties of the elements, hence the fundamental importance of the Periodic table.

As a result, elements which have similar atomic arrangements will have similar properties and will appear at regular intervals thus forming a group or family of elements, e.g. the elements in Group 8 all have a temporarily-filled outside shell of electrons, are all gases, and are all unreactive.

	Atomic no.	K	L	M	N	O	P shell
Helium	2	**2**					
Neon	10	2	**8**				
Argon	18	2	8	**8**			
Krypton	36	2	8	18	**8**		
Xenon	54	2	8	18	18	**8**	
Radon	86	2	8	18	32	18	**8**

The regularity in the structure of the atoms of these elements is obvious.

Lother Meyer, a German scientist, plotted atomic volume against atomic number and discovered that the results corresponded on the whole with the Periodic law of Mendeleeff. Elements with similar properties were found to be in similar positions on his graph, e.g. all the elements at the peaks of the curve possessed similar characteristics.

The modern Periodic table

This consists of eight vertical columns called **groups** and seven horizontal rows called **periods** (see figure 26). Each group consists of elements having the same number of electrons in their outside shells, e.g. sodium has an atomic arrangement of 2,8,1, which means that it is placed in Group 1 of the Periodic table; chlorine has associated with its atom 17 electrons distributed 2,8,7, and therefore finds a place in Group 7. The inert gases with their 'completed' outside shells of 8 (except helium, which

	GROUP																	
PERIOD	1	2											3	4	5	6	7	8
1	1 H																	2 He
2	3 Li	4 Be	TRANSITION ELEMENTS										5 B	6 C	7 N	8 O	9 F	10 Ne
3	11 Na	12 Mg	21 ↓	22 ↓	23 ↓	24 ↓	25 ↓	26 ↓	27 ↓	28 ↓	29 ↓		13 Al	14 Si	15 P	16 S	17 Cl	18 Ar
4	19 K	20 Ca	21 Sc	22 Ti	23 V	24 Cr	25 Mn	26 Fe	27 Co	28 Ni	29 Cu	30 Zn	31 Ga	32 Ge	33 As	34 Se	35 Br	36 Kr
5	37 Rb	38 Sr	39 Y	40 Zr	41 Nb	42 Mo	43 Tc	44 Ru	45 Rh	46 Pd	47 Ag	48 Cd	49 In	50 Sn	51 Sb	52 Te	53 I	54 Xe
6	55 Cs	56 Ba	57 to 71	72 Hf	73 Ta	74 W	75 Re	76 Os	77 Ir	78 Pt	79 Au	80 Hg	81 Tl	82 Pb	83 Bi	84 Po	85 At	86 Rn
7	87 Fr	88 Ra	89 to 103															
				57 La	58 Ce	59 Pr	60 Nd	61 Pm	62 Sm	63 Eu	64 Gd	65 Tb	66 Dy	67 Ho	68 Er	69 Tm	70 Yb	71 Lu
				89 Ac	90 Th	91 Pa	92 U	93 Np	94 Pu	95 Am	96 Cm	97 Bk	98 Cf	99 Es	100 Fm	101 Mv	102 No	103 Lw

Figure 26. The Periodic table

has an outside shell complete with 2) are all placed in Group 8 or Group 0.

Each of the horizontal rows, or periods, contains elements having the same number of shells of electrons, e.g. the inert gas krypton has the electronic arrangement 2, 8, 18, 8, 4 shells of electrons, and will therefore be found in period 4 of the table.

The periods themselves are made up thus:
Period 1 is designated 'very short', containing only the two elements hydrogen and helium.

Periods 2 and 3 each contain 8 elements and are designated 'short periods.'

Periods 4 and 5, designated 'long periods,' each contain 18 elements, 8 of which correspond to the short periods and 10 other elements which are called **transition elements**.

Periods 6 and 7 are also 'long periods' and contain 32 and 17 elements respectively.

General points about the Periodic table

1. The **electropositive nature** of the elements increases according to how far down in their group they appear, e.g. potassium is more electropositive than lithium. The reason for this increase is the increasing size of the atoms, causing the force holding the electrons in the outer shell to become weaker as the distance from the nucleus becomes greater. This is shown by the increase in metallic (basic) character of the elements according to how far down a group they appear, e.g. in Group 4 of the table, lead and tin show many more metallic properties than carbon; likewise, in Group 5, nitrogen – the first element – is a gas, while the final element in the group is the metal bismuth.

2. The electronegativity of the elements increases according to how far across in their period they appear, e.g. Period 3 begins with the highly electropositive element sodium whereas the seventh element is the highly electronegative chlorine. This means that there is an increase in non-metallic (acidic) properties of the elements according to how far across a period they appear.

3. As a result of the trends listed above there are some considerable diagonal similarities between the elements.

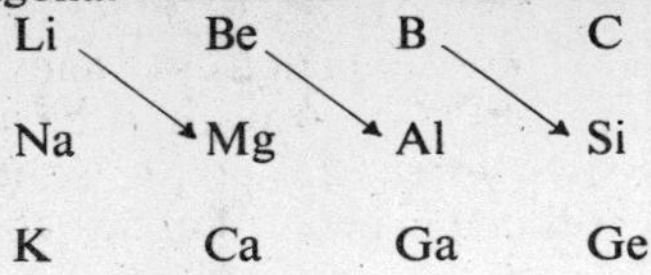

i.e. lithium and magnesium

beryllium and aluminium

boron and silicon.

From the diagonal relationship it follows that the element francium (Fr) is the most electropositive, while fluorine (F) is the most electronegative. Theoretically francium fluoride must be the most stable salt in the universe.

The position of hydrogen

This element has only 1 proton and 1 electron, i.e. it is the only element which does not contain a neutron. It shows properties which are similar to those common to group 1, with 1 electron on its outside shell, and it forms positive ions H^+ similar to Na^+ and K^+.

It also shows similarity to the Group 7 elements in that its physical state is similar to that of fluorine and chlorine. It forms negative ions H^- in hydrides such as NaH, CaH_2, etc., as does Cl^- in NaCl, Br^- in $CaBr_2$. It also forms covalent bonds H – H, H – Cl, etc. Hydrogen is usually considered on its own.

Families of elements

The eight vertical columns in the Periodic table, i.e. the groups, consist of elements which have similar properties and characteristics. These groups, as stated previously, make up families of elements, as opposed to adjacent elements in a period which could be considered 'next-door neighbours' rather than part of the same family. Typical families are to be found in Group 1, the alkali metals and Group 7, the halogens.

The alkali metals are characterised by their reactivity with air and water, all members reacting spontaneously with air to form oxides. As a result they are all stored under oil in airtight containers. They all react violently with water to form very strong **alkalis**, the violence of the reaction varying according to the vertical position of the element in the group (the lower down, the more violent).

Key terms

Periodic table (see figure 26). The elements are shown on this table in order of their increasing atomic numbers so that new horizontal rows (periods) start each time a new outer shell of electrons is started; the vertical columns (groups) consist of elements having the same number of electrons in their respective outside shells.

Periods The horizontal rows of elements in the Periodic table.

Groups The vertical columns in the Periodic table, each consisting of elements having the same number of electrons in their outside shells.

Transition elements These are metals which occupy positions in the Periodic table according to the increase in their respective atomic numbers, but taking into consideration the increase in electron content of their penultimate shells.

Inert elements These, which are all gases, are significant because of their non-reactive nature. They all have 'completed' outer shells of eight electrons, and as such can be placed in Group 8 in the Periodic table.

Alkali metals So called because they all form alkalis when dissolved in water. These elements are found in Group 1 of the Periodic table because they all have just one electron on their outside shell. They constitute a good example of a 'family' of elements, consisting of lithium, sodium, potassium, rubidium and francium.

Halogens Found in Group 7 of the Periodic table, these are another good example of a 'family' of elements, particularly chlorine, bromine and iodine.

Chapter 10
Metals and their Compounds

The elements can be divided into two main categories: metals and non-metals. Frequently, the differences between metals and non-metals are known only in terms of the differences in physical properties. Whilst these are important, there are also chemical differences to bear in mind.

Metals have few electrons in the outer shell of the atom, usually less than four. They form ions by losing electrons and are said to be electropositive, since the ions formed are positive and are deposited at the cathode in electrolysis. The salts of metals are electrovalent and the oxides are also electrovalent and basic (some are also **amphoteric**). Metals usually react with acids, to produce hydrogen.

On the other hand, non-metals have four or more electrons on the outer shell of the atom. They form ions by gaining electrons, and produce negative ions which travel to the anode in electrolysis. The chlorides and oxides are covalent, and the oxides are acidic. Non-metals may be oxidised by concentrated acids, but no hydrogen is released.

Example:	**Metal**, e.g. sodium	**Non-metal**, e.g. sulphur
Electron distribution	2, 8, 1. i.e. 1 electron in outer shell.	2, 8, 6. 6 electrons in outer shell.
Ion formation	$Na - e = Na^+$	$S + 2e \rightarrow S^{2-}$
Electrode reaction	Cathode $Na^+ + e \rightarrow Na$	Anode $S^{2-} - 2e \rightarrow S$
Oxide	(e.g. Na_2O) is electrovalent and basic $Na_2O + H_2O \rightarrow 2NaOH$	(e.g. SO_2) is covalent and acidic $SO_2 + H_2O \rightarrow H_2SO_3$

The essential differences in physical properties are:

	Metals	Non-metals
1.	Usually, high melting and boiling points	Usually, low melting and boiling points
2.	Good conductors of heat and electricity	Poor conductors of heat and electricity
3.	High density	Low density
4.	Malleable and ductile	Brittle
5.	Sonorous	Not sonorous
6.	High lustre	Not usually lustrous
7.	High tensile strength	Low tensile strength

The alkali metals

These are found in Group 1 of the Periodic table and consist of lithium, sodium, potassium, rubidium, caesium and francium. They are called alkali metals because they react with water to form alkalis (soluble bases). They are a very good example of a chemical family of elements. The general properties of the group are exemplified by the metal sodium, which is discussed in detail below.

For examination purposes it is essential to know the general trends and properties of the alkali metals, and have a more detailed knowledge of sodium and potassium and some of their more important compounds.

Sodium (Na)

Occurrence It is found widely in sodium compounds (never as metallic sodium), e.g. as sodium chloride in sea water and salt lakes, as well as in the form of rock salt.

Extraction By the Downs process. Fused (molten) sodium chloride is electrolysed at 600°C (strontium chloride in small amounts is added to lower the melting point). Molten sodium is liberated at the **iron cathode** and chlorine at the **carbon anode.**

$2Na^+ + 2e = 2Na$ iron cathode

$2Cl^- - 2e = Cl_2$ carbon anode

Iron gauze separating the anode and cathode compartments prevents the sodium and chlorine from coming into contact with each other.

Properties of sodium

1. It is a soft shiny metal, easily cut with a knife, which tarnishes immediately on exposure to air.

2. It reacts violently with cold water.
$$2Na + 2H_2O = 2NaOH + H_2\uparrow$$

3. When heated in air it burns with a characteristic yellow flame to form sodium oxide and sodium peroxide.
$$4Na + O_2 = 2Na_2O \quad \text{sodium oxide}$$
$$2Na + O_2 = Na_2O_2 \quad \text{sodium peroxide}$$

4. It will displace hydrogen from dilute acids with explosive violence.
$$2Na + H_2SO_4 = Na_2SO_4 + H_2\uparrow \text{ — molecular}$$
$$2Na + 2H^+ = 2Na^+ + H_2\uparrow \text{ — ionic}$$

Uses of sodium

1. In the manufacture of the 'anti-knock' agent for petrol.

2. In sodium vapour lamps for street lighting, etc.

3. In atomic reactors, as a coolant.

Compounds of sodium

Sodium hydroxide (caustic soda) ($NaOH$)

Manufactured by the electrolysis of concentrated sodium chloride solution using a mercury cathode.
$$NaCl = Na^+ + Cl^-$$

Mercury cathode

$$Na^+ + e = Na$$

The sodium forms an amalgam with mercury which is later decomposed by water.

Carbon anode

$$Cl^- = Cl + e$$
$$2Cl = Cl_2$$

Cl^- ions are discharged in preference to the OH^- ions. Chlorine gas is evolved.

$$2Na(Hg) + 2H_2O = 2NaOH + H_2\uparrow + 2Hg\downarrow$$

The mercury is returned to the cell.

Properties of sodium hydroxide

1. It is a white deliquescent solid.

2. It is an alkali (soluble base) which turns red litmus blue.

3. It forms salts, with acids.
   I.e. Acid + base = salt + water or:
   Acid + alkali = salt + water

4. Being soluble it is used to precipitate insoluble hydroxides from solutions of soluble salts.
   E.g. $FeCl_3 + 3NaOH = 3NaCl + Fe(OH)_3 \downarrow$

5. When heated with ammonium salts ammonia is liberated.
   $NH_4Cl + NaOH = NaCl + H_2O + NH_3 \uparrow$

Uses of sodium hydroxide

In the manufacture of soap, rayon and paper.

Sodium carbonate (Na_2CO_3)

Manufactured by the Kellner-Solvay (ammonia-soda) process. This process depends upon the fact that sodium hydrogen-carbonate (sodium bicarbonate) is not very soluble in brine solution.

The process can be summarised by the equation:

$$NH_4HCO_3 + NaCl = NH_4Cl + NaHCO_3\downarrow$$

Details

Brine is saturated with ammonia and this solution poured into the Solvay tower where it comes into contact with carbon dioxide which is rising up the tower to meet the descending ammoniacal brine. The tower contains baffles which break up the solution, ensuring good mixing and contact with the carbon dioxide. At the bottom of the tower a sludge of sodium hydrogencarbonate collects, owing to its low solubility under the conditions prevailing in the Solvay tower.

$$NH_3 + NaCl + H_2O + CO_2 = NH_4Cl + NaHCO_3 \downarrow$$

Ammoniacal brine

The sodium hydrogencarbonate from the Solvay tower is strongly heated.

$$2NaHCO_3 = Na_2CO_3 + H_2O + CO_2 \uparrow$$

The sodium carbonate is collected, and the carbon dioxide returned to the Solvay tower for further reaction.

The **ammonia** for the process is produced by heating ammonium chloride with slaked lime.

$$Ca(OH)_2 + 2NH_4Cl = CaCl_2 + 2NH_3\uparrow + 2H_2O$$

The slaked lime is produced by adding water to quicklime, which in turn is produced in a lime kiln.

$$CaCO_3 = CaO + CO_2\uparrow$$

$$CaO + H_2O = Ca(OH)_2 \qquad \text{– slaking of lime}$$

The carbon dioxide from this reaction is transferred to the Solvay tower.

This process is a very good example of industrial economics in that the following recoveries of 'raw' materials are made:

1. carbon dioxide from heating sodium hydrogencarbonate;

2. ammonia from ammonium chloride;

$$2NH_4Cl + Ca(OH)_2 \xrightarrow{\text{heat}} CaCl_2 + 2H_2O + 2NH_3\uparrow$$

The calcium hydroxide (slaked lime) is obtained from limestone, carbon dioxide being returned to the tower.

In practice some replenishment of the ammonia is necessary. The only by-product of the reaction is calcium chloride which finds limited use as a desiccant as well as having other limited uses.

Note Potassium hydrogencarbonate cannot be prepared by this process because it is more soluble in brine than sodium hydrogencarbonate and would not be precipitated.

Properties of sodium carbonate

1. It is a white crystalline solid, which in the form of $Na_2CO_3.10H_2O$ is known as washing soda. It is **efflorescent**, i.e. it loses its water of crystallisation on exposure to the air.

2. The hydrated form ($Na_2CO_3.10H_2O$) on heating loses its water of crystallisation to yield anhydrous sodium carbonate, which is stable to heat.

$$Na_2CO_3.10H_2O = Na_2CO_3 + 10H_2O$$

3. It reacts with acids to form a salt, carbon dioxide and water.
   Carbonate + acid = salt + water + carbon dioxide
   $Na_2CO_3 + H_2SO_4 = Na_2SO_4 + H_2O + CO_2\uparrow$

   In this way it neutralises acids, and is used in stomach powders and health salts to neutralise excess stomach acid, the cause of indigestion.

   Note To obtain carbon dioxide from a carbonate or hydrogencarbonate the acid must be in solution. E.g. solid tartaric acid is mixed with sodium carbonate and hydrogencarbonate in some preparations; reaction occurs only when water is added.

4. It is a carbonate precipitator. If a solution of sodium carbonate is added to a solution of a metal salt, other than those of potassium or ammonium, the corresponding carbonate is precipitated.
   E.g. $Pb(NO_3)_2 + Na_2CO_3 = PbCO_3\downarrow + 2NaNO_3$

   $Pb^{2+} + CO_3^{2-} = PbCO_3\downarrow$ — ionic

   Note All carbonates are insoluble except those of sodium, potassium and ammonium.

Uses of sodium carbonate

1. In the manufacture of glass, soap powders and paper.

2. In water-softening.

3. In volumetric analysis.

Sodium hydrogencarbonate (sodium bicarbonate) ($NaHCO_3$)

Manufactured by the ammonia-soda process. It can be prepared in the laboratory by passing carbon dioxide gas into washing soda crystals just covered with water. Fine crystals of the hydrogencarbonate are formed. These are filtered from the mother liquor and **dried in air**.

$$Na_2CO_3 + CO_2 + H_2O = 2NaHCO_3$$

Uses of sodium hydrogencarbonate

1. In health salts and stomach powders.

2. In baking powder as a raising agent.

3. In 'dry' fire extinguishers.

Sodium nitrate (Chile saltpetre) ($NaNO_3$)
Occurs in the natural state.

Properties of sodium nitrate

1. It is a white crystalline solid which is very soluble in water.

2. When strongly heated it melts and eventually decomposes into sodium **nitrite** and oxygen.
   $2NaNO_3 = 2NaNO_2 + O_2$

3. When heated with concentrated sulphuric acid, nitric acid is produced.
   $NaNO_3 + H_2SO_4 = NaHSO_4 + HNO_3$

   This is a good example of a more volatile acid being displaced from its salt by a less or non-volatile acid.

Uses of sodium nitrate

1. In fertilisers (known as nitrate of soda).

2. In the manufacture of nitric acid.

3. In fireworks and flares.

4. In gun powder.

Sodium chloride (common salt) (NaCl)
Occurs naturally as rock salt, and in seas and lakes.

Properties of sodium chloride

1. It is a white crystalline solid which is soluble in water.

2. It is stable to heat.

3. When heated with concentrated sulphuric acid, hydrogen chloride gas (salt gas) is formed. Hydrogen chloride gas dissolves in water to form hydrochloric acid.
   $NaCl + H_2SO_4 = NaHSO_4 + HCl \uparrow$

Another example of displacement of a volatile acid from its salt.

Uses of sodium chloride

1. For seasoning food.
2. As a preservative for butter, meat and fish.
3. In the ammonia-soda process for the manufacture of sodium carbonate.
4. In soap-making (salting out).

Compounds of potassium

These behave in a similar way to compounds of sodium except:

1. they are generally more soluble in water;
2. they are less abundant.

The alkaline earth metals

Found in Group 2 of the Periodic table, the two most important metals for GCE O-level and CSE examination purposes are calcium and magnesium.

Both metals, as would be expected, have two electrons in their outside shells and therefore form the ions Ca^{2+} and Mg^{2+} respectively. They are placed below potassium and sodium in the Activity series.

Calcium (Ca)

Occurrence Calcium does not occur in the native (free) state, but huge quantities of its compounds exist in the form of the carbonate ($CaCO_3$) as limestone, chalk, marble and calcite. E.g. calcium carbonate is found in abundance in its various forms in the White Cliffs of Dover, the Derbyshire hills, Cheddar, etc. It also occurs as dolomite ($CaCO_3MgCO_3$) and anhydrite, which is calcium sulphate ($CaSO_4$). Other common forms include gypsum ($CaSO_42H_2O$) and fluorspar (CaF_2). Calcium compounds are found in solution in natural waters. These compounds are in fact essential to both plants and animals, e.g. bones contain large quantities of calcium phosphate.

Extraction By the electrolysis of fused (molten) calcium chloride to which is added a small amount of calcium fluoride. Calcium forms at the iron cathode, and chlorine at the graphite anode.

$Ca^{2+} + 2e = Ca$ — iron cathode
$2Cl^{-} - 2e = Cl_2$ — graphite anode

Properties of calcium

1. It is a grey, fairly soft metal, quickly tarnishing in air owing to the formation of a crust of calcium oxide (CaO).

2. It reacts with cold water readily, but not violently.
   $Ca + 2H_2O = Ca(OH)_2 + H_2\uparrow$

3. It burns with a bright red flame.
   $2Ca + O_2 = 2CaO$

4. It will displace hydrogen from acids.
   $Ca + 2HNO_3 = Ca(NO_3)_2 + H_2\uparrow$ — molecular
   $Ca + 2H^+ = Ca^{2+} + H_2$ — ionic

Uses of calcium

1. As a reducing agent in the production of the metal thorium.

2. As a deoxidiser in the steel industry.

Compounds of calcium

Calcium oxide (CaO), **quicklime**

Manufactured by roasting limestone in a lime-kiln. The carbon dioxide must be removed for the reaction to proceed to prevent reversal of the reaction.

$CaCO_3 = CaO + CO_2\uparrow$

In the laboratory calcium oxide can be made by strongly heating calcium carbonate for a long time. During the course of this reaction the calcium carbonate (usually in the form of marble) emits an eerie white light which is called 'limelight'. This name comes from the days when stages in theatres were illuminated by heating calcium carbonate.

Properties of calcium oxide

1. It is a fine white powder.

2. It is a basic oxide which reacts with water to form calcium hydroxide (slaked lime). During this reaction great heat is given out (exothermic reaction).
   $CaO + H_2O = Ca(OH)_2$ (solution of limewater)

   The adding of water to calcium oxide (quicklime) is known as the slaking of lime. In excess limewater solution is formed.

3. It reacts with hydrochloric and nitric acids giving the corresponding salts and water.
   $CaO + 2HCl = CaCl_2 + H_2O$

   $CaO + 2HNO_3 = Ca(NO_3)_2 + H_2O$

   The reaction with sulphuric acid is shortlived because of the formation of insoluble calcium sulphate which forms around the calcium oxide and prevents further action.

Uses of calcium oxide

1. In the preparation of slaked lime for the building and agricultural industries.

2. In laboratories for drying gases, e.g. ammonia, which reacts with other drying agents such as concentrated sulphuric acid and calcium chloride.

Calcium hydroxide (slaked lime) ($Ca(OH)_2$)
Manufactured by the action of water on calcium oxide (quicklime). See above.

Properties of calcium hydroxide

1. It is a white amorphous powder which is only slightly soluble in water. In solution it is known as **limewater** which is used to test for carbon dioxide gas. If carbon dioxide is bubbled through limewater for a **short time** the limewater turns milky owing to the formation of **insoluble** calcium carbonate.
   $Ca(OH)_2 + CO_2 = CaCO_3 \downarrow + H_2O$

   If the carbon dioxide is bubbled through limewater for a **long time** (3–4 minutes) the solution becomes clear again

owing to the formation of soluble calcium hydrogencarbonate.
$CaCO_3 + H_2O + CO_2 = Ca(HCO_3)_2$

2. It is a basic hydroxide which reacts with acids to form salts.
E.g. $Ca(OH)_2 + 2HCl = CaCl_2 + 2H_2O$

3. It will liberate ammonia from ammonium salts.
E.g. $2NH_4Cl + Ca(OH)_2 = CaCl_2 + 2NH_3\uparrow + 2H_2O$
$NH_4^+ + OH^- = NH_3\uparrow + H_2O$

4. Solid calcium hydroxide reacts with chlorine to form bleaching powder.
$Ca(OH)_2 + Cl_2 = CaOCl_2 + H_2O$

However if chlorine is bubbled into a concentrated solution of calcium hydroxide, calcium chloride and calcium hypochlorite are formed.
$2OH^- + Cl_2 = Cl^- + OCl^- + H_2O$

Uses of calcium hydroxide
1. In the manufacture of mortar.

2. In agriculture as an additive to the soil, i.e. if the soil is too acid slaked lime can be added to neutralise the acid – yet another example of neutralisation of acids by alkalis or bases.

Calcium carbonate (calcite) ($CaCO_3$)
Occurs naturally as limestone, chalk, and marble. It can be prepared in the laboratory by the action of sodium carbonate solution on calcium chloride solution.
$CaCl_2 + Na_2CO_3 = CaCO_3\downarrow + 2NaCl$
$Ca^{2+} + CO_3^{2-} = CaCO_3\downarrow$

Properties of calcium carbonate
1. It is insoluble in water.

2. It reacts with acids (except sulphuric acid) to form a salt, water and carbon dioxide.
$CaCO_3 + 2HCl = CaCl_2 + H_2O + CO_2\uparrow$

3. It is decomposed by strong heat into calcium oxide and carbon dioxide. This reaction is reversible.
$CaCO_3 \rightleftharpoons CaO + CO_2\uparrow$

Uses of calcium carbonate
1. In the manufacture of cement.

2. In making quicklime in a lime-kiln.

Calcium hydrogencarbonate ($Ca(HCO_3)_2$)
Manufactured by bubbling carbon dioxide gas into limewater for a long time (3–4 minutes – see calcium hydroxide, page 105).

Properties of calcium hydrogencarbonate
1. This compound is responsible for the temporary hardness of water. A solution of calcium hydrogencarbonate is in fact temporary hard water.

2. Calcium hydrogencarbonate exists only in the form of a solution. If it is heated it breaks down into calcium carbonate, water and carbon dioxide.

Calcium sulphate ($CaSO_4$)
Occurs naturally as anhydrite and gypsum, but can be prepared in the laboratory by adding sulphuric acid to a calcium chloride solution.

$$CaCl_2 + H_2SO_4 = CaSO_4\downarrow + 2HCl$$
$$Ca^{2+} + SO_4^{2-} = CaSO_4\downarrow$$

Properties of calcium sulphate
1. It is insoluble in water.

2. If gypsum is heated to 120°–130°C it loses water and forms plaster of Paris, which when mixed with water evolves heat and solidifies to gypsum, expanding slightly. Hence its use for moulds, plaster casts, etc.

Calcium chloride ($CaCl_2$)
Prepared by dissolving calcium carbonate in dilute hydrochloric acid.

$$CaCO_3 + 2HCl = CaCl_2 + CO_2\uparrow + H_2O$$

It is also a major by-product of the ammonia-soda process.

Properties of calcium chloride
1. It is a white porous compound in the anhydrous state.

2. It is very soluble in water.

3. It is very deliquescent.

Uses of calcium chloride
As a desiccant in laboratories, etc.

Calcium nitrate ($Ca(NO_3)_2$)
Occurs in the soil, and can be **manufactured** from limestone and dilute nitric acid.

Properties of calcium nitrate
1. When heated it decomposes into calcium oxide, nitrogen dioxide and oxygen.
$2Ca(NO_3)_2 = 2CaO + 4NO_2\uparrow + O_2\uparrow$

Uses of calcium nitrate
As a fertiliser, sometimes referred to as nitrate of chalk.

Magnesium (Mg)

Occurs in nature in the combined state as magnesium carbonate in magnesite, $MgCO_3$, magnesium sulphate in Epsom salts, $MgSO_4 7H_2O$, as dolomite, which is a double carbonate of magnesium and calcium, etc. Nowadays the sea provides a rich source of magnesium in the form of magnesium chloride.

Extraction By the electrolysis of fused magnesium chloride, using a carbon anode and an iron cathode. Magnesium is formed at the cathode, where in the molten state it rises to the surface of the cell which is covered with natural gas to prevent oxidation. The other product is chlorine which is formed at the anode.

$Mg^{2+} + 2e = Mg$ – iron cathode
$2Cl^- - 2e = Cl_2\uparrow$ – carbon anode.

Properties of magnesium
1. It is a soft silvery-grey metal, stable in dry air but tarnishing rapidly in moist air to form the oxide.

2. It reacts slowly with hot water, but burns in steam if heated to form magnesium hydroxide and hydrogen.
$Mg + 2H_2O = Mg(OH)_2 + H_2\uparrow$

3. It burns in air with a brilliant white light with the formation of the oxide and a small amount of the nitride.
$2Mg + O_2 = 2MgO$
$3Mg + N_2 = Mg_3N_2$

4. It reacts quickly with dilute acids to form a salt and hydrogen.
$Mg + 2HCl = MgCl_2 + H_2 \uparrow$
$Mg + 2H^+ = Mg^{2+} + H_2 \uparrow$

Note Magnesium will displace hydrogen from very dilute nitric acid (1%). More concentrated acid gives the oxides of nitrogen.

5. It will burn in carbon dioxide, sulphur dioxide, and nitrous oxide to form the oxide, releasing the non-metals.
$2Mg + CO_2 = 2MgO + C$
$2Mg + SO_2 = 2MgO + S$
$Mg + N_2O = MgO + N_2$

6. It will reduce most metallic oxides on heating.
$3Mg + Al_2O_3 = 2Al + 3MgO$

7. It combines directly with non-metals when heated.
$Mg + S = MgS$
$Mg + Cl_2 = MgCl_2$

Uses of magnesium

1. In flashlight photography, flares and shells.

2. In the production of light alloys, particularly with aluminium.

3. (To a limited extent) in the production of spheroidal graphite (S.G.) iron.

4. In synthetic organic chemistry for Grignard reagents and reactions.

5. To protect ships' hulls and underground cables by the process known as cathodic protection.

6. As a reducing agent in the extraction of some metals, e.g. titanium.

Compounds of magnesium

Magnesium oxide (MgO)
Prepared by heating magnesium in oxygen,

$$2Mg + O_2 = 2MgO$$

or by heating magnesium carbonate, hydroxide or nitrate.

$$MgCO_3 = MgO + CO_2\uparrow$$
$$Mg(OH)_2 = MgO + H_2O$$
$$2Mg(NO_3)_2 = 2MgO + O_2\uparrow + 4NO_2\uparrow$$

Properties of magnesium oxide

1. It is a fine white powder.

2. It is only slightly soluble in water, but still produces a fairly alkaline solution.
$MgO + H_2O = Mg(OH)_2$

3. As a basic oxide it reacts with dilute acids to form salts.
$MgO + 2HCl = MgCl_2 + H_2O$

Uses of magnesium oxide

1. As a refractory lining for furnaces, crucibles, etc.

2. In medicines, to cure acidity of the stomach.

Magnesium hydroxide ($Mg(OH)_2$)
Prepared by adding sodium hydroxide solution to a solution of magnesium salts. Double decomposition occurs with the precipitation of magnesium hydroxide.

$$MgCl_2 + 2NaOH = Mg(OH)_2\downarrow + 2NaCl$$

Properties of magnesium hydroxide
It is a normal basic hydroxide, and reacts with dilute acids to give salts.

$$Mg(OH)_2 + H_2SO_4 = MgSO_4 + 2H_2O$$

Uses of magnesium hydroxide
In the sugar industry for the extraction of sugar from molasses.

Magnesium carbonate ($MgCO_3$)
Occurs naturally associated with calcium carbonate as dolomite ($MgCO_3CaCO_3$) and also as magnesite ($MgCO_3$). In the pure state it is a white solid which when heated decomposes easily.

$$MgCO_3 = MgO + CO_2\uparrow$$

Magnesium sulphate ($MgSO_4$)

Occurrence Crystalline magnesium sulphate ($MgSO_4 7H_2O$) is called Epsom salts because it occurs abundantly in the natural springs of Epsom. A white crystalline solid, it is used in medicines, dyeing, soap and paint. (This salt is prepared in the usual way for metal sulphates.)

Aluminium (Al)

Occurrence Approximately 7 per cent of the earth's crust is aluminium, the most widely distributed metal in nature. It is contained in most clays and rocks combined as the silicate though extraction from these sources is not as yet economically viable. However, it can be extracted from the mineral **bauxite** ($Al_2O_3 2H_2O$) which is impure aluminium oxide. Cryolite (Na_3AlF_6) is another source of the metal.

Extraction Aluminium is extracted from the mineral bauxite, which is first purified. The crushed bauxite is treated with caustic soda solution. It dissolves, forming sodium aluminate, and any insoluble matter, such as iron (II) oxide and silica, is filtered off. Aluminium hydroxide is then precipitated from the remaining solution of sodium aluminate by adding freshly precipitated aluminium hydroxide to seed the solution.

$$\left.\begin{array}{l} Al_2O_3 + 2NaOH \rightleftharpoons \underset{\text{sodium aluminate}}{2NaAlO_2} + H_2O \\ NaAlO_2 + 2H_2O = NaOH + Al(OH)_3 \end{array}\right\} \text{Bayer process}$$

The aluminium hydroxide is filtered off and strongly heated forming pure Al_2O_3.

The alumina is mixed with a small amount of cryolite (Na_3AlF_6) and calcium fluoride to lower the melting point and increase the fluidity of the cell feed. This mixture is then electrolysed at 600°C using carbon electrodes. Molten aluminium formed at the cathode collects at the bottom of the cell from where it is syphoned off. Oxygen which is liberated at the anodes gradually burns them away, which makes regular replacement of these electrodes necessary.

$$\begin{array}{llll} Al^{3+} + 3e & = Al & - & \text{carbon cathode} \\ 2O^{2-} - 4e & = O_2 & - & \text{carbon anode} \\ (C + O_2 & = CO_2 & - & \text{anode burning}) \\ 4Al^{3+} + 6O^{2-} & = 4Al + 3O_2 \uparrow & & \text{total reaction} \end{array}$$

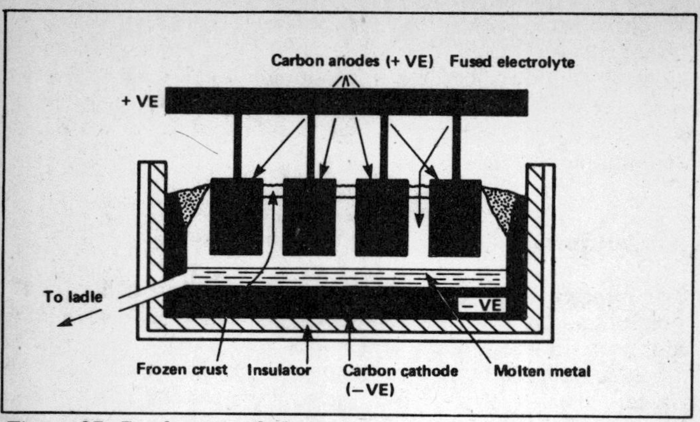

Figure 27. Production of aluminium

Properties of aluminium

1. It is a silvery white metal of low density.

2. A reactive metal, quickly forming a tenacious self-healing oxide layer in air which prevents further reaction. (The process called **anodising** involves the thickening of this oxide film.)

3. It does not react with nitric acid under any conditions (passive), nor with dilute sulphuric acid. However, it will react with concentrated sulphuric acid to give aluminium sulphate and sulphur dioxide and with hydrochloric acid to give hydrogen and aluminium chloride.
   $2Al + 6H_2SO_4 = Al_2(SO_4)_3 + 3SO_2\uparrow + 6H_2O$
   $2Al + 6HCl = 2AlCl_3 + 3H_2\uparrow$

4. It will burn in air if heated to 1,000°C to form aluminium oxide and nitride.
   $4Al + 3O_2 = 2Al_2O_3$
   $2Al + N_2 = 2AlN$

5. It reacts vigorously with caustic soda solution to give sodium aluminate and hydrogen.
   $2Al + 2NaOH + 2H_2O = 2NaAlO_2 + 3H_2\uparrow$

6. It reacts with non-metals on heating.
   $2Al + 3Cl_2 = 2AlCl_3$
   $2Al + 3S = Al_2S_3$

7. At high temperatures it will react with the oxides of iron, manganese and chromium, reducing them to the metals. The reaction with iron is known as the Thermit process, and is very exothermic; for this reason it is sometimes used to weld iron.
$Fe_2O_3 + 2Al = Al_2O_3 + 2Fe - \Delta H$

Uses of aluminium

1. In alloys for electricity cables and aircraft. (The most common alloy is one with 4% copper, widely known as **duralumin**.)
2. In cooking utensils and paint.
3. In the kitchen in the form of foil for cooking, wrapping, etc., for it can be rolled into very thin foil. Aluminium is one of the few metals with a surface that can be dyed.

Compounds of aluminium

Aluminium oxide (Al_2O_3)
This is a white insoluble powder prepared by the action of heat on aluminium hydroxide.

$$2Al(OH)_3 = Al_2O_3 + 3H_2O$$

It is an amphoteric oxide, which means that in the presence of a base it acts as an acid while with an acid it would act as a base.

E.g.	$Al_2O_3 + 2NaOH$	$= 2NaAlO_2$	$+ H_2O$
	acid + alkali	= salt	+ water
	$Al_2O_3 + 3H_2SO_4$	$= Al_2(SO_4)_3$	$+ 3H_2O$
	base + acid	= salt	+ water

Aluminium hydroxide ($Al(OH)_3$)
This is hydrated aluminium oxide and can be prepared by the action of ammonium hydroxide on a solution of an aluminium salt.

$$AlCl_3 + 3NH_4OH = Al(OH)_3\downarrow + 3NH_4Cl$$

Note If sodium or potassium hydroxides are added to a solution of an aluminium salt, aluminium hydroxide would be precipitated but owing to its amphoteric nature it would dissolve in excess of the alkali to form the aluminate.

$$AlCl_3 + 3NaOH = Al(OH)_3\downarrow + 3NaCl$$
$$Al(OH)_3 + NaOH = NaAlO_2 + 2H_2O$$

Aluminium chloride ($AlCl_3$)

Prepared by heating aluminium in a stream of dry chlorine gas.

$$2Al + 3Cl_2 = 2AlCl_3$$

This salt, like all other aluminium compounds, is white.

Zinc (Zn)

Occurrence This metal occurs as the sulphide zinc blende ZnS in addition to various other combined forms, e.g. calamine ($ZnCO_3$), zincite (ZnO), etc.

Extraction By the reduction by carbon of zinc oxide. Remember that in terms of the Activity series zinc is the first metal that can be extracted by reduction of its oxide. Zinc **sulphide** is roasted to convert it into the oxide.

$$2ZnS + 3O_2 = 2ZnO + 2SO_2\uparrow$$

The sulphur dioxide produced is used in the manufacture of sulphuric acid. The oxide is then heated with carbon (in the form of coke) when it is reduced to zinc.

$$ZnO + C = Zn + CO\uparrow$$

Zinc has a comparatively low boiling point (907°C) and the vapour can be distilled from it, cooled and collected.

Properties of zinc

1. It is a hard, grey, white metal.

2. It is stable in air, but forms a protective layer of the carbonate in damp air.

3. If heated it burns in air to form zinc oxide.
   $$2Zn + O_2 = 2ZnO$$

4. It decomposes steam when heated in it.
   $$Zn + H_2O = ZnO + H_2\uparrow$$

5. With dilute and concentrated hydrochloric acid and dilute sulphuric acid hydrogen is evolved. This is a convenient method of preparing hydrogen in a laboratory.
   $$Zn + 2HCl = ZnCl_2 + H_2\uparrow$$
   $$Zn + H_2SO_4 = ZnSO_4 + H_2\uparrow$$

With concentrated sulphuric acid a variety of products is obtained. Similarly with nitric acid a variety of products is also obtained, depending upon the concentration and temperature of the acid.

6. It reacts with hot aqueous sodium hydroxide to give sodium zincate and hydrogen.

$$Zn + 2NaOH = \underset{\text{sodium zincate}}{Na_2ZnO_2} + H_2\uparrow$$

7. It will react with non-metals when heated in their presence. E.g. $Zn + Cl_2 = ZnCl_2$

8. Being a strongly electropositive metal, it will displace other metals from solutions of their salts.

E.g. $Zn + CuSO_4 = ZnSO_4 + Cu\downarrow$

$Zn + Cu^{2+} = Zn^{2+} + Cu\downarrow$

Uses of zinc

1. In the production of **galvanised** steel products, e.g. steel sheets, dustbins, etc.
2. When alloyed with copper it produces a series of alloys known as **brasses,** e.g. 70% copper and 30% zinc is known as cartridge brass.
3. In large quantities in the manufacture of batteries and certain utensils.

Compounds of zinc

Zinc oxide

Prepared by heating either zinc nitrate, carbonate or hydroxide.
E.g. $2Zn(NO_3)_2 = 2ZnO + 4NO_2\uparrow + O_2\uparrow$

It can be reduced by heating with an excess of carbon.

$$ZnO + C = Zn + CO\uparrow$$

Zinc oxide is amphoteric, forming zinc salts with acids,

$$\underset{\text{basic oxide}}{ZnO} + \underset{\text{acid}}{H_2SO_4} = \underset{\text{salt}}{ZnSO_4} + \underset{\text{water}}{H_2O}$$

and **zincates** with alkalis.

$$2NaOH + ZnO \quad = Na_2ZnO_2 + H_2O$$
$$\underset{\text{alkali}}{\downarrow} \quad + \quad \underset{\text{acidic oxide}}{\downarrow}$$

When heated zinc oxide turns yellow, but reverts to its original white colour when cold, i.e. it is yellow when hot and white when cold.

Zinc hydroxide ($Zn(OH)_2$)
Prepared by adding sufficient sodium hydroxide solution to a solution of a zinc salt.

$$ZnSO_4 + 2NaOH = Zn(OH)_2 \downarrow + Na_2SO_4$$

The precipitate will redissolve in excess alkali to form the zincate.

$$Zn(OH)_2 + 2NaOH = \underset{\text{sodium zincate}}{2Na_2ZnO_2} + 2H_2O$$

Zinc carbonate
Prepared by adding sodium carbonate solution to a zinc salt solution.

$$ZnSO_4 + Na_2CO_3 = ZnCO_3 \downarrow + Na_2SO_4$$

The salt will decompose on heating.

$$ZnCO_3 = ZnO + CO_2 \uparrow$$

Zinc chloride
Prepared by heating zinc metal in a current of chlorine gas.

$$Zn + Cl_2 = ZnCl_2$$

All zinc compounds are white in colour.

Iron (Fe)

Occurrence As haematite (Fe_2O_3), magnetite (Fe_3O_4), iron pyrites (FeS_2), etc. Iron is the second most widely distributed metal in nature constituting some 4 per cent of the earth's crust.

Extraction Occupies a middle position in the Activity series. iron (III) oxide can be reduced by carbon in the form of coke. This process takes place in a blast furnace, so called because of the blast of hot air fed in at the bottom of the furnace. The raw materials used in the extraction of iron from its ores in the blast furnace are air, iron ore, limestone and coke. This is another good example of an industrial process. The reactions are as follows.

1. The coke burns to form carbon monoxide and carbon dioxide. The carbon dioxide reacts further with more coke to produce carbon monoxide.
   $2C + O_2 = 2CO\uparrow$
   $C + O_2 = CO_2\uparrow$
   $CO_2 + C = 2CO\uparrow$

2. The carbon monoxide reduces the iron ore to iron, which collects at the bottom of the furnace ready to be tapped.
   $Fe_2O_3 + 3CO \rightleftharpoons 2Fe + 3CO_2\uparrow$

3. The limestone decomposes.
   $CaCO_3 \rightleftharpoons CaO + CO_2\uparrow$

   The carbon dioxide produced then reacts with coke to form carbon monoxide which in turn reduces more iron ore.

4. The quicklime (CaO) produced reacts with the siliceous material in the ore to form a slag which floats on the top of the molten iron from where it can be tapped.
   $CaO + SiO_2 = CaSiO_3$
   The calcium oxide also reacts with phosphorus impurities and these are also removed as part of the slag.

The iron produced in the furnace is called **pig iron** and contains 3–5 per cent of carbon together with phosphorus, manganese, silicon, sulphur, and other impurities. The pig iron can be converted into steel, which is basically an iron-carbon alloy, by reducing the amount of carbon and other impurities by oxidation. Steel can contain up to 1.5 per cent carbon, combined with iron in the form of cementite (Fe_3C). The properties and type of steel produced depend upon the amount of carbon dissolved, e.g. a steel containing 0.1–0.2 per cent carbon is called **mild steel** and could be used for pressings in the car industry because of its comparative malleability and ductility. A steel that contained 0.8 per cent carbon would be very hard, and useful for the manufacture of tools, hence the name **tool steels**. Carbon is the hardening element in steels.

Most steels nowadays are produced by the Linz-Donawitz (or L.D.) process. This process requires oxygen of high purity to be blown at supersonic speed on to the surface of pig iron contained in a vessel by means of a vertical pipe or lance inserted through the mouth of the vessel, so that the impurities are oxidised away. This process replaces the much slower open-hearth furnace method. Special steels such as the many stainless steels and electrical steels are produced in electric furnaces.

Properties of iron

1. It is a grey, magnetic metal having a melting point of 1535°C.

2. It is stable in dry air but if heated will form ferrosoferric oxide (Fe_3O_4), the black oxide of iron.

3. It will react with steam if heated strongly.
   $3Fe + 4H_2O \rightleftharpoons Fe_3O_4 + 4H_2 \uparrow$

4. It will react with dilute hydrochloric and sulphuric acids to liberate hydrogen.
   $2Fe + 6HCl = 2FeCl_3 + 3H_2 \uparrow$
   $2Fe + 3H_2SO_4 = Fe_2(SO_4)_3 + 3H_2 \uparrow$

   Concentrated nitric acid renders iron passive.

5. It reacts with non-metals when heated to form sulphides and chlorides, etc.
   $Fe + S = FeS$ iron (II) sulphide
   $Fe + Cl_2 = FeCl_2$ iron (II) chloride

Rusting of iron

In the presence of air and water (moist air), iron will rust to form hydrated iron (III) oxide ($Fe_2O_3.3H_2O$). When rusting, iron is merely reverting back to its natural state. It is found naturally as the oxide, from which man extracts the metal. The natural environment in which man exists is an oxidising one. When exposed the iron reacts with oxygen in the air to form iron oxide. Rusting is therefore a natural phenomenon. Most methods used for the prevention of rusting set out to prevent the metal coming into contact with the corroding medium. These methods involve coating the metal, e.g. by painting, oiling, greasing, tinning, plating, galvanising or plastic coating.

Experiment to show that iron needs both water and air (oxygen) to rust

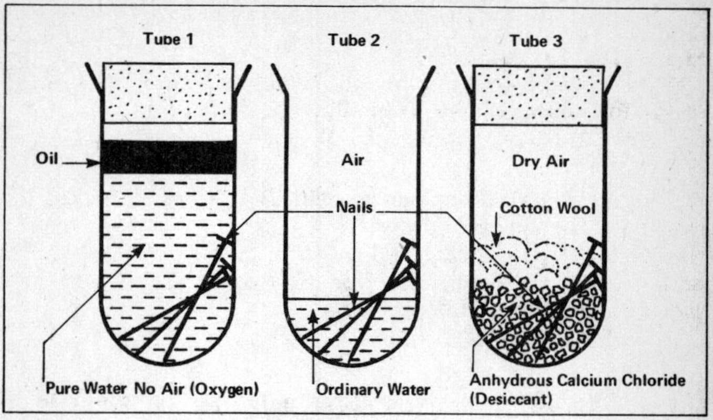

Figure 28. Experiment on rusting

Test tube 1 containing pure water, i.e. the water must be air-(oxygen-) free. The nails show no sign of rusting.

Test tube 2 containing both air and water. The nails have rusted.

Test tube 3 containing dry air. No rusting is evident.

Uses of iron

These are legion, but the most important is steelmaking. One only has to envisage a world without iron and steel to realise how vital and important this element is.

Compounds of iron

Iron can form two series of compounds, those which contain the Fe^{2+} ion (iron II or ferrous ion), and those containing the Fe^{3+} ion (iron III or ferric ion).

Iron (II) compounds All iron (II) compounds are reducing agents because they can be oxidised to iron (III) compounds.

$$Fe^{2+} - e = Fe^{3+}$$

Solutions of iron (II) salts are easily oxidised by the air to iron (III) salts, e.g. if a fresh solution of iron (II) sulphate is made, the watery green solution will begin to go brown very quickly when exposed to the air.

Iron (II) sulphate ($FeSO_4$)
This is a watery-green-coloured crystalline solid, prepared by dissolving iron wire in **air-free** dilute sulphuric acid.

$$Fe + H_2SO_4 = FeSO_4 + H_2 \uparrow$$

The solution crystallises out as $FeSO_4 7H_2O$. When heated the salt at first becomes anhydrous,

$$FeSO_4 7H_2O = FeSO_4 + 7H_2O$$

and then breaks down into iron (III) oxide (Fe_2O_3), sulphur dioxide and sulphur trioxide.

$$2FeSO_4 = Fe_2O_3 + SO_2 \uparrow + SO_3 \uparrow$$

Note Iron (II) sulphate is used in testing for a nitrate in solution. A freshly prepared solution of iron (II) sulphate is mixed with the suspected nitrate and concentrated sulphuric acid added very slowly and carefully. If a nitrate is present the compound **nitroso-iron (II) sulphate** ($FeSO_4NO$) is formed as a brown ring in the test tube. The test is called the **brown ring test** (see page 160).

Iron (II) chloride ($FeCl_2$)
Like iron (II) sulphate this compound is prepared by the action of iron with the appropriate acid, i.e. dilute hydrochloric acid.

$$Fe + 2HCl = FeCl_2 + H_2 \uparrow$$

This is a pale green compound which can easily be oxidised to the more familiar reddish-brown iron (III) chloride.

Iron (II) hydroxide ($Fe(OH)_2$)
Prepared by adding sodium hydroxide solution to an iron (II) salt solution.

$$FeSO_4 + 2NaOH = Fe(OH)_2 \downarrow + Na_2SO_4$$

It is precipitated as a flocculent pale green solid. It is a basic oxide, reacting with acids to form salts, and also acting as a reducing agent.

Iron (II) sulphide (FeS)

Prepared by heating iron filings and sulphur, great heat being evolved in the process.

$$Fe + S = FeS \text{ (a salt of hydrogen sulphide)}$$

It is a black insoluble substance which in large masses has a metallic lustre. Used in the production of hydrogen sulphide by reaction with dilute acids.

$$FeS + 2HCl = FeCl_2 + H_2S \uparrow \quad \text{(preparation of hydrogen sulphide)}$$

Iron (III) compounds

All iron (III) compounds are oxidising agents because they can be reduced to the iron (II) ion state.

$$Fe^{3+} + e = Fe^{2+}$$

The majority of these compounds are dark yellow to reddish-brown compared with the green coloration of the iron (II) compounds. They are very much more stable than the iron (II) compounds.

Iron (III) oxide (Fe_2O_3)

This is a dark red compound which is found naturally but can be prepared by heating iron (II) sulphate strongly.

$$2FeSO_4 = Fe_2O_3 + SO_2 \uparrow + SO_3 \uparrow$$

This compound is mainly basic in reaction and reacts with acids to form salts.

E.g. $Fe_2O_3 + 6HCl = 2FeCl_3 + 3H_2O$

It will also react with bases to form ferrites

$$2NaOH + Fe_2O_3 = 2NaFeO_2 + H_2O$$

i.e. it is amphoteric. It can be reduced to iron by heating with hydrogen, carbon or carbon monoxide.

Ferrosoferric oxide (Fe_3O_4)

This is a black compound which occurs naturally as magnetite. It can be prepared by heating iron in steam.

$$3Fe + 4H_2O \rightleftharpoons Fe_3O_4 + 4H_2 \uparrow$$

In its reaction it can be regarded as acting as a mixture of FeO and

Fe_2O_3, thus upon reacting it would yield mixtures of iron (II) and iron (III) compounds. It is a mixed oxide.

$$Fe_3O_4 + 8HCl = FeCl_2 + 2FeCl_3 + 4H_2O$$

Note Iron (II) oxide (FeO) is a very unstable compound.

Iron (III) hydroxide ($Fe(OH)_3$)
This is a reddish-brown compound, prepared by adding sodium hydroxide solution to a solution of an iron (III) salt.

$$FeCl_3 + 3NaOH = Fe(OH)_3\downarrow + 3NaCl$$

It is a weak base and will react with acids.

Iron (III) sulphate ($Fe_2(SO_4)_3$)
This is a yellowish compound prepared by heating iron (II) sulphate in concentrated sulphuric acid.

$$2FeSO_4 + 2H_2SO_4 = Fe_2(SO_4)_3 + SO_2\uparrow + 2H_2O$$

Iron (III) chloride ($FeCl_3$)
This is a yellow-brown compound prepared by heating iron wire in chlorine.

$$2Fe + 3Cl_2 = 2FeCl_3$$

Remember if the higher chloride is required, iron is reacted with chlorine. However, if the lower chloride is wanted, iron is reacted with hydrochloric acid.

$$2Fe + 3Cl_2 = 2FeCl_3 \quad \text{(iron III)}$$
$$Fe + 2HCl = FeCl_2 + H_2\uparrow \quad \text{(iron II)}$$

Tests for iron (II) and iron (III) compounds

1. **Colour** Iron (II) compounds are generally green, while iron (III) compounds are dark yellow to brown.
2. If an alkali (NaOH or KOH) is added to a solution of an **iron (II) salt,** iron (II) hydroxide which is **green** is precipitated. In the case of the **iron (III)** compound a **reddish-brown** precipitate of iron (III) hydroxide is formed.
3. The addition of potassium hexacyanoferrate (II) solution to an aqueous solution of iron (II) ions yields a white precipitate which rapidly changes to blue.
4. The addition of potassium hexacyanoferrate (III) solution to an aqueous solution of iron (II) ions produces a dark blue precipitate.
   Note Potassium hexacyanoferrate (II) solution used to be

called potassium ferrocyanide while potassium hexacyano-ferrate (III) solution used to be called potassium ferricyanide.

Lead (Pb)

Occurrence Native lead has been found in small quantities. The main ore is galena-lead sulphide (PbS). It also occurs as the carbonate and sulphate.

Extraction The sulphide ore is first roasted to convert it to the oxide.

$2PbS + 3O_2 = 2PbO + 2SO_2 \uparrow$

The lead (II) oxide formed is then heated with carbon in the form of coke which reduces the oxide to the metal.

$PbO + C = Pb + CO \uparrow$ (carbon monoxide)

and $PbO + CO = Pb + CO_2 \uparrow$

Properties of lead

1. It is a soft, heavy, grey metal, so soft that it can be rolled into thin sheets with great ease.

2. It tarnishes slowly in air, but the oxide film produced protects it from further reaction. If heated in air, lead will oxidise very slowly to form lead (II) oxide (PbO) and on further roasting red lead oxide (Pb_3O_4) is formed.

3. It has very little reaction with water or steam.

4. It reacts with dilute nitric acid to form lead nitrate and nitric oxide (NO).
   $3Pb + 8HNO_3 = 3Pb(NO_3)_2 + 2NO \uparrow + 4H_2O$
   Its reaction with other acids is very slow, as would be expected in view of its position in the Activity series.

5. It reacts with chlorine to form lead (II) chloride.
   $Pb + Cl_2 = PbCl_2$

Uses of lead

1. In radiation shields and batteries.
2. Alloys, e.g. solder, which is an alloy of lead and tin.
3. In lead base bearing metals when alloyed with antimony and tin.

4. In the manufacture of anti-knock additives for petrol.

Compounds of lead

Lead is another metal with a variable valency, which means it can form lead (II) and sometimes lead (IV) compounds.

Lead (II) compounds

Lead (II) oxide (PbO)

This is a heavy yellow compound sometimes called **litharge.** It can be prepared in the laboratory by heating any lead salt except the chloride.

E.g. $2Pb(NO_3)_2 = 2PbO + 4NO_2 + O_2\uparrow$

$PbCO_3 = PbO + CO_2\uparrow$

$2Pb_3O_4 = 6PbO + O_2\uparrow$

The compound is very easily reduced to the metal by heating with carbon. It will react with acids to form salts.

E.g. $PbO + 2HNO_3 = Pb(NO_3)_2 + H_2O$

and will dissolve in sodium hydroxide to form sodium plumbite, i.e. it is feebly amphoteric.

$$PbO + 2NaOH = Na_2PbO_2 + H_2O$$

Remember that an amphoteric oxide reacts with both acids and bases to form salts.

Lead (II) hydroxide ($Pb(OH)_2$)

Prepared by the action of an alkali on a solution of a lead salt. (The only soluble salt of lead is lead nitrate.) A white precipitate of lead hydroxide is formed which will dissolve in excess alkali to form the plumbite (see lead (II) oxide).

$$Pb(NO_3)_2 + 2NaOH = Pb(OH)_2\downarrow + 2NaNO_3$$

Lead (II) nitrate $Pb(NO_3)_2$

This is the only soluble salt of lead. A white crystalline solid which is prepared by the action of dilute nitric acid on lead, the oxide or the carbonate.

$$PbO + 2HNO_3 = Pb(NO_3)_2 + H_2O$$

When heated, lead (II) nitrate breaks down into lead (II) oxide, nitrogen dioxide and oxygen.

$$2Pb(NO_3)_2 = 2PbO + 4NO_2\uparrow + O_2\uparrow$$

It is interesting that when heated the compound crackles (decrepitates) and no water of crystallisation is evolved. Despite its crystalline character, lead (II) nitrate does not contain any water of crystallisation.

Lead (II) chloride ($PbCl_2$)
This is a heavy white powder, prepared by the direct action of chlorine on lead (synthesis).

$$Pb + Cl_2 = PbCl_2$$

Insoluble, it can also be prepared by double decomposition of a soluble lead salt and a soluble chloride.

$$Pb(NO_3)_2 + 2HCl = PbCl_2\downarrow + 2HNO_3$$

It dissolves in hot water, and resolidifies on cooling, i.e. though soluble in hot water, it is virtually insoluble in cold.

Lead (IV) compounds
The only compound relevant to the GCE O-level and CSE examinations is **lead (IV) oxide,** a dark brown solid which is insoluble in water. It is prepared by heating nitric acid with **red lead oxide** Pb_3O_4. 'Red lead' is a bright red powder and can be regarded as a **mixed oxide** of lead (II) oxide and lead (IV) oxide, i.e. red lead oxide Pb_3O_4 is made up of $2PbO, PbO_2$. This can be demonstrated by its action with nitric acid.

$$Pb_3O_4 + 4HNO_3 = PbO_2\downarrow + 2Pb(NO_3)_2 + 2H_2O$$

The reaction can be split up as follows:

$$\underset{\text{(mixed oxide)}}{Pb_3O_4}\ \begin{cases} 2PbO + 4HNO_3 = 2Pb(NO_3)_2 + 2H_2O \\ PbO_2\text{: remains unchanged} \end{cases}$$

When lead (IV) oxide is heated with concentrated hydrochloric acid, it oxidises the acid to chlorine.

$$PbO_2 + 4HCl = PbCl_2 + Cl_2\uparrow + 2H_2O$$

Copper (Cu)

Occurrence Copper occurs in the native state, and as copper pyrites ($CuFeS_2$), malachite ($CuCO_3Cu(OH)_2$), etc.

Extraction From copper pyrites. First the ore is roasted in air.

$$2CuFeS_2 + 4O_2 = Cu_2S + 2FeO + 3SO_2\uparrow$$

The products are then heated with sand which removes the iron as slag.

$$FeO + SiO_2 = FeSiO_3 \text{ (slag)}$$

The fused copper (I) sulphide (the matte) is heated in a converter and air is blown in.

$$2Cu_2S + 3O_2 = 2Cu_2O + 2SO_2 \uparrow$$
$$Cu_2S + 2Cu_2O = 6Cu + SO_2 \uparrow$$

The molten copper is cast and is purified by electrolysis, using copper (II) sulphate as the electrolyte.

Properties of copper

1. It is a reddish-brown metal and an excellent conductor of heat and electricity.

2. In damp air it forms a green layer of the basic carbonate, **verdigris ($CuCO_3Cu(OH)_2$).**

3. It has no reaction with water, being lower than hydrogen in the Activity series. It is an unreactive metal.

4. As stated above, it is less electropositive than hydrogen and will not displace hydrogen from dilute acids. When heated with concentrated sulphuric acid, copper sulphate and sulphur dioxide are formed.

$$\begin{aligned} Cu + H_2SO_4 &= \cancel{CuO} + H_2O + SO_2 \uparrow \\ \cancel{CuO} + H_2SO_4 &= CuSO_4 + H_2O \\ \hline Cu + 2H_2SO_4 &= CuSO_4 + 2H_2O + SO_2 \uparrow \end{aligned}$$

In this reaction the sulphuric acid first oxidises the copper to copper (II) oxide, which then reacts further to form a salt and water, i.e. sulphuric acid acts first as an oxidising agent then as an acid.

Nitric acid will react with copper to give copper (II) nitrate and nitrogen oxides, e.g. with moderately concentrated nitric acid:

$$3Cu + 8HNO_3 = 3Cu(NO_3)_2 + 2NO \uparrow + 4H_2O$$

with cold concentrated nitric acid:

$$Cu + 4HNO_3 = Cu(NO_3)_2 + 2NO_2 + 2H_2O$$

Uses of copper

1. In coinage. Alloyed with tin, bronze is formed from which the present bronze currency of the UK is made; the 50p and 10p coins are made from a copper-nickel alloy.
2. Alloyed with zinc, a whole range of alloys called **brasses** are produced.
3. Being a good conductor of heat and electricity, copper is used for hot-water pipes and cisterns, and electrical wiring and contacts.
4. Copper salts are used extensively in agriculture and horticulture to combat the harmful effects of some insects, fungi etc. on plants and crops.

Compounds of copper

Copper can form two ions, copper (I) Cu^+ (cuprous) and copper (II) Cu^{2+} (cupric). The only stable copper (I) compounds are insoluble because in solution copper (I) compounds decompose into copper (II) compounds and copper.

Copper (I) compounds
The only relevant one is copper (I) oxide (Cu_2O). This is a brick-red powder, prepared by reducing Fehling's solution (a mixture of copper (II) sulphate, sodium hydroxide, and sodium potassium tartrate) with glucose. It is insoluble in water but will react with acids to form copper (I) salts.

Copper (II) compounds
These are stable copper compounds, usually green or blue in colour.

Copper (II) oxide (CuO)
This is a black insoluble powder, prepared by heating copper nitrate, carbonate or hydroxide.

$$2Cu(NO_3)_2 = 2CuO + 4NO_2\uparrow + O_2\uparrow$$
$$CuCO_3 = CuO + CO_2\uparrow$$
$$Cu(OH)_2 = CuO + H_2O$$

Copper (II) oxide is a base dissolving in acids to form salts.

$$CuO + 2HCl = CuCl_2 + H_2O$$

It is readily reduced to copper by heating in hydrogen or carbon.

$$CuO + H_2 = Cu + H_2O$$

Copper (II) hydroxide ($Cu(OH)_2$)
Prepared by adding sodium hydroxide solution to a copper (II) sulphate solution; a blue precipitate of the hydroxide is formed.

$$CuSO_4 + 2NaOH = Cu(OH)_2 \downarrow + Na_2SO_4$$

Note If ammonium hydroxide were used instead of sodium hydroxide, the blue copper (II) hydroxide precipitate formed would redissolve in excess ammonia solution to form the dark blue complex tetrammine copper (II) ion. Copper (II) hydroxide decomposes on heating.

$$Cu(OH)_2 = CuO + H_2O$$

It is a base and reacts with acids to form salts.

$$Cu(OH)_2 + H_2SO_4 = CuSO_4 + 2H_2O$$

Copper (II) sulphate ($CuSO_45H_2O$)
This a blue crystalline solid, which can be prepared by dissolving copper (II) oxide in dilute sulphuric acid. When heated the water of crystallisation is lost, rendering the salt anhydrous. Anhydrous copper (II) sulphate is a white powder. Water or liquids containing water will restore the blue colour if added to anhydrous copper sulphate. This is a common test for detecting the presence of water.

$$\underset{\text{blue}}{CuSO_45H_2O} = \underset{\text{white}}{CuSO_4} + 5H_2O$$

Copper (II) nitrate ($Cu(NO_3)_2$)
This is a royal blue crystalline solid which is deliquescent. Prepared by dissolving copper (II) oxide in dilute nitric acid.

$$CuO + 2HNO_3 = Cu(NO_3)_2 + H_2O$$

It is very soluble in water and forms crystals of $Cu(NO_3)_23H_2O$ which decompose on heating.

$$2Cu(NO_3)_2 = 2CuO + 4NO_2 \uparrow + O_2 \uparrow$$

Copper (II) carbonate ($CuCO_3$)
This is a green insoluble powder which is not known in a pure state. It breaks down very readily on heating.

$$CuCO_3 = CuO + CO_2 \uparrow$$

Chapter 11
Non-metals and their Compounds

Oxygen (O_2)

Prepared on an industrial scale by the fractional distillation of air, from which oxygen and nitrogen are separated by physical means (proof that air is a mixture). The gas is prepared in the laboratory by:

1. heating mercury (II) oxide, lead (IV) oxide or red lead oxide;
   $2HgO = 2Hg + O_2 \uparrow$, $2PbO_2 = 2PbO + O_2 \uparrow$,
   $2Pb_3O_4 = 6PbO + O_2 \uparrow$

2. heating potassium chlorate mixed with manganese (IV) oxide, which acts as a catalyst causing the potassium chlorate to yield oxygen at a lower temperature than when heated alone;
   $2KClO_3 = 2KCl + 3O_2 \uparrow$

3. dropping '10' volume hydrogen peroxide on to manganese (IV) oxide (catalyst) in the cold;
   $2H_2O_2 = 2H_2O + O_2 \uparrow$ (catalytic decomposition)

4. electrolysis of acidulated water.

In the laboratory, method 3 is the most popular. Being only slightly soluble in water, oxygen can be collected over water.

Properties of oxygen

It is a colourless, odourless gas, slightly heavier than air. Oxygen is slightly soluble in water supporting fish and plant life. Although it will not burn (non-combustible) it will allow things to burn in it (supporter of combustion) much more readily than in air. When elements burn in oxygen, oxides are formed and the element is said to be oxidised.

E.g. $C + O_2 = CO_2 \uparrow$ $2Mg + O_2 = 2MgO$

Note Oxides of metals are **basic**, while oxides of non-metals are **acidic**. Other kinds of oxides are:

1. **amphoteric oxides**, which act as bases with acids and as acids in the presence of bases.
   $ZnO + 2HCl = ZnCl_2 + H_2O$ (ZnO as a base)
   $ZnO + 2NaOH = Na_2ZnO_2 + H_2O$ (ZnO as an acid)

2. **neutral oxides**, like carbon monoxide and water, which show neither acidic nor basic properties.

3. **peroxides**, which are metallic oxides containing more oxygen than the corresponding basic oxide. **All** give hydrogen peroxide (H_2O_2) with dilute acids.

$$Na_2O_2 + H_2SO_4 = H_2O_2 + Na_2SO_4$$

Hydrogen peroxide is prepared by treating a peroxide (usually barium peroxide) with dilute sulphuric acid.

$$BaO_2 + H_2SO_4 = H_2O_2 + BaSO_4 \downarrow$$

It is a colourless liquid which decomposes easily when exposed to light, warmed, or treated with manganese (IV) oxide.

$$2H_2O_2 = 2H_2O + O_2 \uparrow$$

Test for oxygen It will rekindle a glowing splint.

Uses of oxygen

1. For producing steel from pig iron in the L.D. converter (see page 118).
2. In oxyacetylene welding.
3. For breathing apparatus.

Air is composed of 78 per cent nitrogen, 21 per cent oxygen, 1 per cent inert gases, 0.03 per cent carbon dioxide together with variable amounts of water vapour and impurities.

Air as a mixture

1. The composition of air is almost constant but not absolutely so.

2. The properties of air are exactly what would be predicted knowing its composition, e.g. oxygen supports combustion readily, nitrogen does not support combustion, so air is a reasonably good supporter of combustion. If air were a compound then by definition its properties would be quite different from those of its constituent elements.

3. Air can be separated into its component parts quite easily, e.g. by liquefaction then evaporation.

4. When the constituent gases of air are mixed in the correct proportions, the resulting mixture has all the properties of air, yet there is no evidence of any chemical reaction having

occurred, i.e. no heat is produced and no change in volume occurs.

Analysis of air

Some water, together with a few crystals of pyrogallol and a pellet of sodium hydroxide, is added to a long graduated tube sealed at one end. Pyrogallol absorbs oxygen and sodium hydroxide absorbs carbon dioxide. The other end of the tube is immediately closed with a rubber bung. The volume of air which fills the rest of the tube is known. The tube is shaken for fifteen minutes, then the bung is removed under water. Water rushes in to fill the space previously taken by oxygen and carbon dioxide. The volume of nitrogen (and noble gases) left is noted after first adjusting the tube so that the height of the liquid inside and outside the tube is the same. Thus the volume of nitrogen and the total volume of air are known and hence the percentage of nitrogen in the air can be calculated.

5. Air as a raw material

Air is a most important chemical in the industrial preparation of oxygen, ammonia and nitric acid. It is also essential in the burning of fuels.

Note In the vital processes of **breathing** and **burning** approximately one fifth of the air is used up, i.e. the oxygen. The process of **rusting** also uses up one fifth of the air.

6. Pollution of the air

Since the upsurge of modern industry and the invention of the internal combustion engine, the air in and around large towns and cities has gradually become more and more polluted. Waste gases from industry and machinery have been allowed to escape into the air, and even the ordinary domestic 'open' coal-fire produces large quantities of gases. Air pollution contributed greatly to the once-frequent London smogs, and has damaged many buildings. Carbon monoxide (CO) and sulphur dioxide (SO_2) rank very high in the list of gases which pollute the atmosphere. Legislation has been introduced to lessen air pollution e.g. by using smokeless fuels, and filtering and absorbing harmful waste gases from industry. In the USA efforts have been made to introduce a system to deal with motor-car exhaust fumes; however this has so far proved too costly to be viable. The prevention of air pollution and the harm it does to animal- and plant-life takes high priority among research scientists today.

Hydrogen (H_2)

Preparation

1. Treat the Z.I.M. metals (zinc, iron, magnesium) with dilute hydrochloric or sulphuric acid.
   $Zn + H_2SO_4 = ZnSO_4 + H_2$

   This is a convenient laboratory preparation. The gas can be collected over water, by the downward displacement of air (hydrogen is lighter than air) or upward delivery.

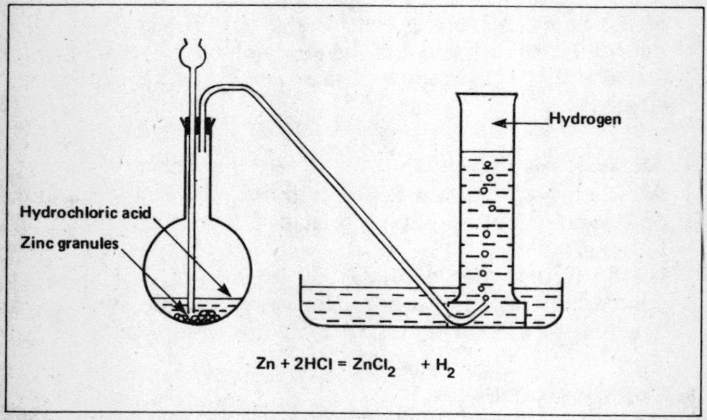

Figure 29. Preparation of hydrogen

(**Note** *If a gas is required dry, never dry it and then collect it over water.*)

2. Potassium, sodium and calcium will displace hydrogen from cold water. Potassium and sodium do so with explosive violence. In each case the corresponding alkalis are formed
   $2K + 2H_2O = 2KOH + H_2 \uparrow$

3. Magnesium, aluminium and iron will, when heated, displace hydrogen from steam.
   $Mg + H_2O = MgO + H_2 \uparrow$

4. Acidulated water can be split into hydrogen and oxygen by electrolysis.

Properties of hydrogen

1. It is a colourless, odourless gas almost insoluble in water. It is lighter than air (lightest of all the elements).

2. It is combustible.

   $2H_2 + O_2 = 2H_2O$ (burns with blue flame)

   $H_2 + Cl_2 = 2HCl$ (photochemical change)

3. It does not support combustion.

4. It combines directly with sulphur.

   $H_2 + S = H_2S \uparrow$ (synthesis of hydrogen sulphide)

5. It is a very powerful reducing agent, reducing the oxides of the L.I.C. (lead, iron, copper) metals. One of many examples is:

   $CuO + H_2 = Cu + H_2O$

Test for hydrogen With air the gas forms an explosive mixture. If this is ignited it explodes ('pops' in small quantities).

Important facts relating to hydrogen

1. Before igniting a hydrogen supply, a sample of the gas should be collected and tested to ensure that it burns quietly (contains no air).

2. The reducing power of the gas can be illustrated by the reduction of metal oxides, e.g. copper (II) oxide, lead (II) oxide, etc. In each case the metal is formed as a powder, while drops of liquid collect which can be shown to be water. A knowledge of this experiment is useful in that questions on the following points can be answered.
   (a) To show that hydrogen is a reducing agent.
   (b) In the case of metals low in the Activity series, to prepare such metals from their oxides.
   (c) To illustrate the gravimetric composition of water, and the law of constant composition.

Uses of hydrogen

1. In the synthesis of ammonia in the **Haber process.**
2. In the manufacture of margarine, and many organic chemicals.
3. For the oxy-hydrogen flame.

Water (H_2O)

Produced naturally, but in many chemical reactions, among them the following, water is a product:

1. in the reduction of the oxides of the L.I.C. (lead, iron, copper) metals;
   $PbO + H_2 = Pb + H_2O$ (etc.)
2. when hydrogen is burned in air or oxygen, water is formed;
   $2H_2 + O_2 = 2H_2O$ (synthesis of water)

Tests for water

1. Water or liquids containing water will turn anhydrous copper(II) sulphate blue, and/or blue cobalt chloride paper pink.

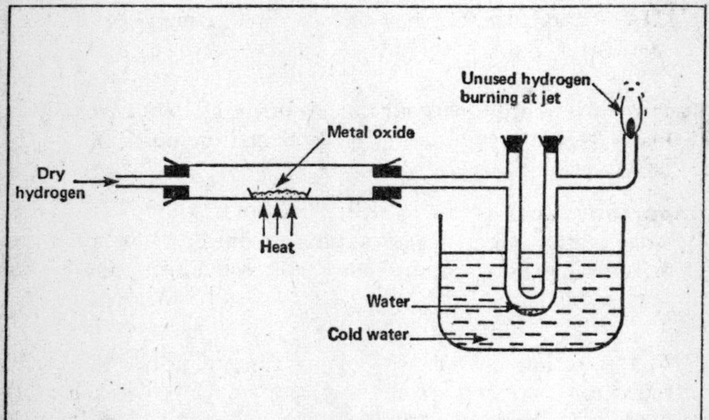

Figure 30. Reduction of an oxide by hydrogen

2. Pure water at sea level has a boiling point of 100°C and a freezing point of 0°C.

Properties of water

1. It is a colourless, odourless liquid.
2. It is a neutral oxide.
3. It will react with metals high in the Activity series to form hydrogen.
   $2K + 2H_2O = 2KOH + H_2 \uparrow$ (cold water)
   $3Fe + 4H_2O \rightleftharpoons Fe_2O_4 + 4H_2 \uparrow$ (steam)

4. With **acid anhydrides** it forms acids.
   $SO_3 + H_2O = H_2SO_4$

5. It forms true bases with **basic anhydrides.**
   $NH_3 + H_2O = NH_4OH$

6. An electric current will split water into oxygen and hydrogen.

7. It is an excellent solvent, as well as being one of the major supporters of life.

Water of crystallisation

This is the water which is usually chemically combined with some substances when they crystallise from an aqueous solution. A salt which contains water of crystallisation is said to be hydrated. Examples of hydrated salts are: washing soda ($Na_2CO_3 10H_2O$); magnesium sulphate ($MgSO_4 7H_2O$); copper (II) sulphate ($CuSO_4 5H_2O$). It is the water of crystallisation that gives crystals their crystalline form and sometimes their colour. Some salts are never hydrated, e.g. lead (II) nitrate $Pb(NO_3)_2$.

Deliquescence

When a substance absorbs moisture from the air and dissolves in it to form a solution it is said to be deliquescent, e.g. sodium hydroxide (NaOH). A substance such as concentrated sulphuric acid which absorbs moisture from the air without undergoing a change in state is said to be hygroscopic. Copper (II) oxide acquires moisture from the atmosphere but does not change its state. Some hygroscopic substances, e.g. silica gel, are used in desiccators for the drying of materials or storing in a dry atmosphere.

Efflorescence

When a substance loses its water of crystallisation on exposure to the air it is said to be efflorescent, e.g. washing soda ($Na_2CO_3.10H_2O$). The crystals acquire a white powdery surface of anhydrous sodium carbonate and eventually the crystal crumbles to a powder.

Dehydrating agents

A dehydrating agent will remove the elements of water from pure and perfectly dry compounds, e.g. concentrated sulphuric acid converts blue copper (II) sulphate crystals to white anhydrous copper (II) sulphate.

Drying agents

A drying agent will absorb water or water vapour from other substances. It is not to be confused with a dehydrating agent.

E.g.

Drying agent	Gas which may be dried
Calcium chloride	Hydrogen
Calcium oxide	Ammonia
Concentrated sulphuric acid	Chlorine

Pollution of water

Surveys have shown that up to 30 per cent of the river water in the United Kingdom is polluted in some way. In certain coastal areas the sea has been visibly polluted to a considerable extent by untreated sewage, oil and waste materials. Poisons and radioactive materials are deposited in water in and around the country. Much thought and major replanning are needed, together with stricter legislation, to prevent the disaster that will inevitably result if the situation continues unchanged. To combat pollution, many waste-disposal plants have been constructed throughout the country, new techniques introduced and research directed towards methods of cleaning up existing polluted waters and maintaining their relative cleanliness in the future.

Despite these measures there is still much to be done before this evergrowing problem is solved, especially now that 'sea and water farming' is playing a greater role than ever in the fight against world malnutrition. Water may be the saviour of this planet and the people who live on it, provided it fulfils the purpose for which it was intended.

Hardness of water

Hard water is water which will not easily lather with soap. It is caused by dissolved calcium or magnesium compounds (usually sulphates or hydrogencarbonates) producing Ca^{2+} or Mg^{2+} ions.

Temporary hardness of water is hardness that **can** be removed by boiling and is caused by the presence of dissolved hydrogencarbonates of calcium and magnesium.

Permanent hardness of water is hardness that **cannot** be removed by boiling and is caused by the presence of the chlorides and/or the sulphates of calcium and magnesium.

Natural formation of hard water

1. Rain water containing carbon dioxide forms hydrogencarbonates on flowing over soil containing calcium or magnesium carbonates.

$$\underset{\text{soil and rocks}}{CaCO_3} + \underbrace{H_2O + CO_2}_{\text{rain water}} = Ca(HCO_3)_2$$

2. It is also formed by rain flowing over rocks containing calcium or magnesium compounds, e.g. calcium sulphate ($CaSO_4$).

Formation in the laboratory

If carbon dioxide is bubbled into a solution of calcium hydroxide (limewater), the limewater first turns milky because of the formation of insoluble calcium carbonate.

$$Ca(OH)_2 + CO_2 = CaCO_3\downarrow + H_2O$$

If the passing of carbon dioxide through the solution is continued it will eventually clear owing to the formation of soluble calcium hydrogencarbonate solution, which is a solution of temporary hard water.

$$CaCO_3 + H_2O + CO_2 = Ca(HCO_3)_2$$

Note Limewater will become milky in the presence of carbon dioxide, a phenomenon which serves as the well-known test for carbon dioxide. Remember that it can equally well serve as a test for limewater.

Removal of hardness in water

Any method employed depends on removing the calcium or magnesium ions from the water and converting them into insoluble compounds.

Removal of temporary hardness

1. By **boiling,** when the soluble hydrogencarbonate is decomposed forming insoluble carbonate and is therefore removed from the solution.

$$Ca(HCO_3)_2 = CaCO_3\downarrow + CO_2\uparrow + H_2O$$

2. **By Clark's process**, which depends on the addition of a calculated quantity of slaked lime.

$$Ca(HCO_3)_2 + Ca(OH)_2 = 2CaCO_3\downarrow + 2H_2O$$

Removal of temporary and permanent hardness

1. By **distillation**, when all the soluble salts are left behind.

This process is far too expensive to be used on an industrial scale.

2. **By adding washing soda** ($Na_2CO_3.10H_2O$), which causes the soluble hydrogencarbonate to become an insoluble carbonate by double decomposition.

$$Ca(HCO_3)_2 + Na_2CO_3 = 2NaHCO_3 + CaCO_3\downarrow$$

3. **By the Permutit process** 'Permutit' is a trade name for sodium aluminium silicate. When hard water runs over it an ion exchange takes place, the permutit exchanging its Na^+ ions for the Ca^{2+} or Mg^{2+} ions present in the water. Thus insoluble calcium or magnesium aluminium silicate is formed, and the water softened.

$$Ca(HCO_3)_2 + \underset{\text{permutit}}{2Na\mathbf{P}} = 2NaHCO_3 + Ca\mathbf{P}_2$$

The permutit can be regenerated by standing overnight in concentrated brine solution.

$$2NaCl + Ca\mathbf{P}_2 = CaCl_2 + 2Na\mathbf{P}$$

4. **By adding excess soap** The soap (sodium stearate) reacts with the water to form insoluble calcium stearate, which usually floats on the top of the water as a scum.

$$CaSO_4 + \underset{\text{soap}}{2NaSt} = Na_2SO_4 + \underset{\text{scum}}{CaSt\downarrow} \text{ (St – Stearate)}$$

The relative hardness of different waters can be compared by adding soap solution to equal volumes of each sample of water until a permanent lather is obtained.

25 cm^3 of distilled water are titrated with soap solution until a 'permanent' lather results, i.e. the lather lasts for two minutes. This operation is repeated for each water sample to be tested (equal volumes), and the volume of soap solution used in each case is compared with that required for distilled water.

Carbon (C)

This is an **allotropic** element, which means it can exist in two or more distinct physical forms – diamond and graphite. Other forms of carbon such as animal charcoal, lamp black, sugar charcoal, have been shown by X-ray analysis to be finely divided graphite.

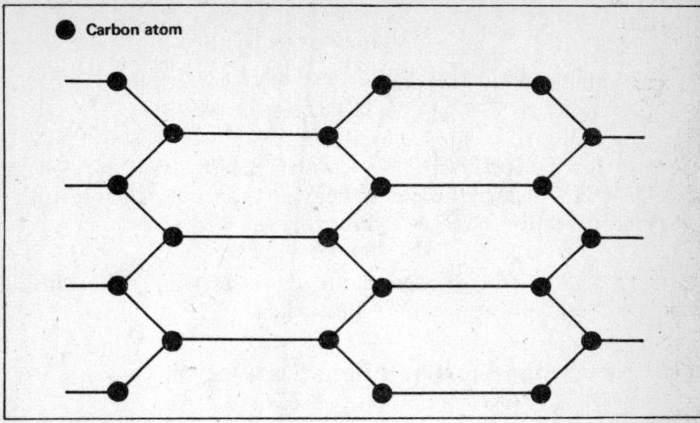

Figure 31. Structure of graphite

Uses of carbon allotropes

Diamond In jewellery; in industry, where its great hardness makes it invaluable for cutting, etc.

Graphite As a lubricant; in pencil 'leads'; in the manufacture of electrodes.

Allotropy

Allotropy, or the ability of an element to exist in two or more distinct physical forms, arises when atoms of the same element are arranged in different ways, giving rise to differences in physical properties (see diamond, figure 22, and graphite, figure 31). Uses should be correlated with properties and structures.

Carbon dioxide

Laboratory preparation

1. Treat any carbonate or hydrogencarbonate with any dilute acid. (There are few exceptions.)

$$CaCO_3 + 2HCl = CaCl_2 + H_2CO_3\ (H_2O + CO_2)$$
$$NaHCO_3 + HCl = NaCl + H_2CO_3\ (H_2O + CO_2)$$

Note the similarity with the preparation of sulphur dioxide.

$$Na_2\mathbf{SO}_3 + 2HCl = 2NaCl + H_2\mathbf{SO}_3\ (H_2O + \mathbf{SO}_2)$$
$$Na_2\mathbf{CO}_3 + 2HCl = 2NaCl + H_2\mathbf{CO}_3\ (H_2O + \mathbf{CO}_2)$$

2. Heat any carbonate except those of sodium or potassium.

   $ZnCO_3 = ZnO + CO_2\uparrow$

   Remember that zinc oxide is yellow when hot and white when cold.

3. Heat any hydrogencarbonate. There are no exceptions.

   $2NaHCO_3 = Na_2CO_3 + H_2CO_3\ (H_2O + CO_2)$

Carbon dioxide is collected by downward delivery or the upward displacement of air (the gas is heavier than air). Compare this with the collection of a gas lighter than air (figure 34).

Carbon dioxide is produced on an industrial scale during alcoholic fermentation, and in lime-kilns.

Test for carbon dioxide It turns limewater milky.

$Ca(OH)_2 + CO_2 = CaCO_3\downarrow + H_2O$

Properties of carbon dioxide

1. It is a colourless, odourless gas which is heavier than air, easily solidified (−80°C) to form 'dry ice'.

2. It is soluble in water producing a solution of the weak, unstable carbonic acid.

   $H_2O + CO_2 = H_2CO_3$ (carbonic acid)

3. It dissolves in alkalis to form carbonates and hydrogencarbonates.

   $2NaOH + CO_2 = Na_2CO_3 + H_2O$ (carbonate)

   $Na_2CO_3 + H_2O + CO_2 = 2NaHCO_3$ (hydrogencarbonate)

4. It does not support combustion unless the substance is burning very strongly, e.g. magnesium.

   $CO_2 + 2Mg = 2MgO + C$ (black and white ashes)

5. It turns limewater milky because of the precipitation of calcium carbonate, and then clear again with the formation of soluble calcium hydrogencarbonate.

   $Ca(OH)_2 + CO_2 = CaCO_3\downarrow + H_2O$ (milky)

   $CaCO_3 + H_2O + CO_2 = Ca(HCO_3)_2$ (clear)

   Remember that all hydrogencarbonates decompose on heating therefore they cannot be formed in hot solutions.

6. It is reduced to carbon monoxide by carbon (see page 117):
   $CO_2 + C = 2CO\uparrow$.

Uses of carbon dioxide

1. In the manufacture of mineral waters. The gas is dissolved under pressure, and when the pressure is released there is considerable effervescence (fizzing).
2. Solid carbon dioxide is used as a refrigerant ('dry ice').
3. Being much heavier than air it can be used in fire extinguishers.
4. Baking powder contains sodium hydrogencarbonate, which when heated liberates carbon dioxide, which acts as a raising agent.

Sulphur (S)

Occurs as elementary sulphur (native) in Sicily and the USA. It also occurs in sulphides, e.g. zinc blende (ZnS) and as sulphates, e.g. anhydrite ($CaSO_4$) and gypsum ($CaSO_4 2H_2O$).

Extraction

1. In Sicily it is dug up as sulphur rock and purified by melting on sloping hearths; as the sulphur flows down it leaves behind the impurities.

2. By the Frasch process (USA), used when the sulphur lies underground below quicksands and cannot be mined conventionally.

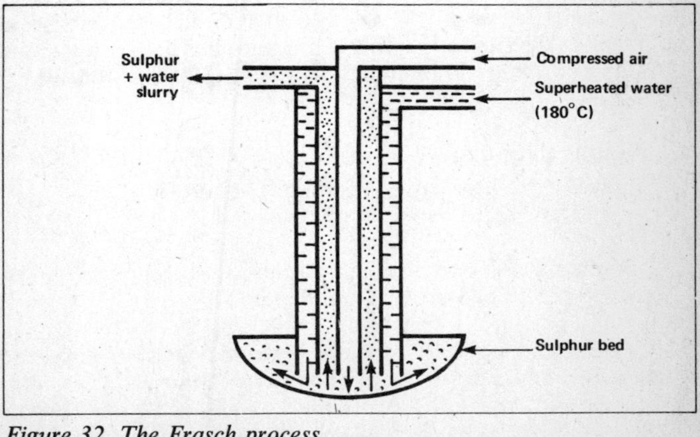

Figure 32. The Frasch process

Three concentric pipes are sunk into the ground. Water superheated to 180°C is forced under pressure down the outer pipe, causing the sulphur to melt. A blast of hot compressed air is blown down the inner pipe, forcing the liquid sulphur up the middle tube. The liquid sulphur is run into vats where it solidifies. No further purification is necessary as the sulphur is about 99.5 per cent pure (figure 32).

The allotropic forms of sulphur

Sulphur is another element that exhibits allotropy.

1. **Monoclinic (prismatic)** This is obtained by melting sulphur and allowing the liquid to cool. Immediately a crust forms on the surface, two holes are pierced in the crust and any liquid sulphur poured off. Needle-like crystals, long thin prisms, are seen beneath the crust.

2. **Rhombic (octahedral)** This is prepared by dissolving powdered roll sulphur in carbon disulphide (CS_2). Any undissolved sulphur is filtered off and the solution left to evaporate. Octahedral crystals of rhombic sulphur are formed. **Note** Rhombic sulphur is the most stable form below 96°C, and other allotropes gradually revert to it. 96°C is the transition point.

Action of heat on sulphur

Sulphur melts at 112°C forming an amber-coloured liquid. As the temperature rises the colour darkens until it is almost black at the boiling point of 444°C. On first melting it becomes viscous (toffee-like). On further heating it becomes less and less viscous. If boiling sulphur is poured into water **plastic sulphur** is formed.

Properties of sulphur

1. It burns with a blue flame to form sulphur dioxide.
   $S + O_2 = SO_2$

2. It combines with many metals on heating to produce sulphides.
   E.g. $Fe + S = FeS$ (iron (II) sulphide)

3. It has no action with any acid except **hot concentrated sulphuric acid** or **hot concentrated nitric acid**.
   $S + 2H_2SO_4 = 3SO_2\uparrow + 2H_2O$
   This reaction can be explained as follows:

$$\left.\begin{array}{r} 2H_2O \\ S + 2SO_2 \\ 2O \end{array}\right\} 2H_2SO_4 = 3SO_2\uparrow + 2H_2O$$

The sulphur is oxidised to sulphur dioxide while the sulphuric acid is reduced to sulphur dioxide. Hot concentrated sulphuric acid oxidises sulphur to sulphur **tri**oxide (SO_3) which then forms sulphuric acid.

$$S + 6HNO_3 = SO_3 + 3H_2O + 6NO_2\uparrow$$

$$\left.\begin{array}{r} 3H_2O \\ S + 6NO_2 \\ 3O \end{array}\right\} = \begin{array}{l} H_2SO_4 \\ (SO_3 + H_2O) \end{array} \quad + 2H_2O + 6NO_2\uparrow$$

i.e. $S + 6HNO_3 = H_2SO_4 + 2H_2O + 6NO_2\uparrow$

The sulphur is oxidised to sulphur trioxide while the nitric acid is reduced to nitrogen dioxide (NO_2)

Note $2H_2SO_4$ can for convenience be written as $2H_2O$, $2SO_2$, $2O$; likewise $6HNO_3$ can be written as $3H_2O$, $6NO_2$, $3O$. These forms render the equations easier to understand.

Uses of sulphur

1. In vulcanising rubber.
2. In the manufacture of sulphuric acid and calcium hydrogen-sulphite (used in the paper industry).
3. As an agent for spraying hop gardens and vineyards.
4. In drugs, many of which contain sulphur.

Sulphur dioxide (SO_2)

Preparation

1. Burn sulphur in air or oxygen.
   $S + O_2 = SO_2$ (not a good preparation)

2. By the action of **hot concentrated** sulphuric acid on any common metal. Copper is usually used. The copper is first oxidised by the acid to copper (II) oxide, and thence copper (II) sulphate, the sulphuric acid being reduced to sulphur dioxide.

$$Cu + H_2SO_4 = CuO + H_2O + SO_2\uparrow \quad -(1)$$
$$CuO + H_2SO_4 = CuSO_4 + H_2O \quad -(2)$$
$$\text{Adding, } Cu + 2H_2SO_4 = CuSO_4 + 2H_2O + SO_2\uparrow$$

In equation (1) the sulphuric acid is acting as an oxidising agent, while in equation (2) it is acting as an acid. The residue in the flask is grey, being made up of anhydrous copper (II) sulphate together with black copper (I) sulphide (Cu_2S)

produced by a side reaction. To obtain copper (II) sulphate crystals pour the residue carefully into water, filter, and allow to crystallise out.

3. Treat any sulphite or hydrogensulphite with dilute hydrochloric or sulphuric acid.
   E.g.
   $Na_2SO_3 + 2HCl = 2NaCl + H_2SO_3\ (H_2O + SO_2)$ or:
   $NaHSO_3 + HCl = NaCl + H_2SO_3\ (H_2O + SO_2)$ compare:
   $Na_2CO_3 + 2HCl = 2NaCl + H_2CO_3\ (H_2O + CO_2)$ or:
   $NaHCO_3 + HCl = NaCl + H_2CO_3\ (H_2O + CO_2)$
   Note It is important to observe the similarity between the preparations of CO_2 and SO_2.

4. Heat any hydrogensulphite.
   E.g.
   $2NaHSO_3 = Na_2SO_3 + H_2SO_3\ (H_2O + SO_2)$ compare:
   $2NaHCO_3 = Na_2CO_3 + H_2CO_3\ (H_2O + CO_2)$

5. Commercially, by burning sulphide ores.
   E.g.
   $2ZnS + 3O_2 = 2ZnO + 2SO_2\uparrow$
   $4FeS_2 + 11O_2 = 2Fe_2O_3 + 8SO_2\uparrow$

6. From anhydrite ($CaSO_4$). The anhydrite is ground with coke, sand and alumina (Al_2O_3) and the mixture heated to about 1400°C. The anhydrite is reduced by the coke.
   $2CaSO_4 + C = 2CaO + 2SO_2\uparrow + CO_2\uparrow$

Collection of sulphur dioxide

Sulphur dioxide, being heavier than air, is collected by the upward displacement of air.

If dry gas is required it is first passed through concentrated sulphuric acid and collected over mercury.

Tests for sulphur dioxide

1. It has a very distinctive smell.

2. It turns filter paper soaked in potassium dichromate from orange to green (so does hydrogen sulphide, but sulphur dioxide, unlike hydrogen sulphide, does not blacken lead acetate paper).

3. It bleaches moist litmus paper.

Properties of sulphur dioxide

1. It is a colourless gas with a characteristic taste and smell.

2. It is much heavier than air.

3. It is easily liquefied and is used in refrigeration. (Most laboratories possess a syphon of sulphur dioxide.)

4. It is soluble in water, forming a solution of the weak acid sulphurous acid. Sulphur dioxide is said to be the **anhydride** of sulphurous acid.
   $SO_2 + H_2O = H_2SO_3$ (unstable) compare:
   $CO_2 + H_2O = H_2CO_3$ (unstable)

5. It dissolves in alkalis, forming sulphites then hydrogensulphites (compare with carbon dioxide).
   $SO_2 + 2NaOH = Na_2SO_3 + H_2O$ (sulphite) then:
   $Na_2SO_3 + H_2O + SO_2 = 2NaHSO_3$ (hydrogensulphite)
   Like hydrogencarbonates, hydrogensulphites cannot be formed in hot solution.

6. It does not burn or support combustion unless the substance is burning **very** strongly, e.g. magnesium. (Once again, note the similarity with carbon dioxide.)
   $SO_2 + 2Mg = 2MgO + S$ (yellow deposit)
   In the above reaction the sulphur dioxide is acting as an **oxidising** agent, the magnesium being oxidised to magnesium oxide while the sulphur dioxide is reduced to sulphur.

Sulphur dioxide as a reducing agent

In the presence of water, sulphur dioxide dissolves to form sulphurous acid (H_2SO_3), which is a very good reducing agent in that it can be easily oxidised.

$$H_2SO_3 + \text{'O'} = H_2SO_4$$

1. Therefore sulphur dioxide is a mild bleaching agent in the presence of water because the sulphurous acid takes oxygen away from the dye.
   H_2SO_3 + dye = H_2SO_4 + (dye – O)
   coloured colourless

2. Sulphur dioxide will decolorise an acid solution of potassium permanganate ($KMnO_4$) by reduction.
$2KMnO_4 + 5SO_2 + 2H_2O = K_2SO_4 + 2MnSO_4 + 2H_2SO_4$
purple colourless
Similarly it will turn a solution of potassium dichromate from orange to green.
$K_2Cr_2O_7 + 3SO_2 + H_2SO_4 = K_2SO_4 + Cr_2(SO_4)_3 + H_2O$
orange green

Sulphur dioxide acts as a reducing agent with chlorine gas in the presence of water.
$Cl_2 + H_2O = HCl + HClO$ ($HClO = HCl +$ 'O')
$SO_2 + H_2O = H_2SO_3$ nascent oxygen
$Cl_2 + SO_2 + 2H_2O = H_2SO_4 + 2HCl$

Sulphur dioxide as an oxidising agent

1. It forms a yellow cloud of sulphur when it reacts with hydrogen sulphide.
$2H_2S + SO_2 = 2H_2O + 3S$

2. It will allow magnesium to burn in it, forming magnesium oxide and sulphur.
$SO_2 + 2Mg = 2MgO + S$ (yellow deposit)

Sulphurous acid

Sulphurous acid (H_2SO_3) is a weak unstable acid (rather like carbonic acid, H_2CO_3). It forms salts called sulphites, e.g. sodium sulphite (Na_2SO_3) and, as it contains two replaceable hydrogen atoms (dibasic), it also forms hydrogensulphites, e.g. sodium hydrogensulphite ($NaHSO_3$).

Distinction between sulphates and sulphites

1. Both sulphates and sulphites in solution give a white precipitate with barium chloride ($BaCl_2$) solution, producing barium sulphate ($BaSO_4$) and barium sulphite ($BaSO_3$) respectively, both of which are insoluble in water. However, in the case of the sulphite, the precipitate disappears when dilute hydrochloric acid is added, while the sulphate precipitate is permanent.

2. When warmed with dilute hydrochloric acid, sulphites produce sulphur dioxide while sulphates do not.

Uses of sulphur dioxide

1. For bleaching wood pulp in the paper industry.
2. In the manufacture of sulphuric acid.
3. In the preservation of foodstuffs.

Sulphur trioxide (SO_3)

This is often referred to as the acid anhydride of sulphuric acid, i.e. sulphuric acid without water.

E.g. $H_2SO_4 = H_2O + SO_3$

Preparation

1. The laboratory method is a model of the industrial process called the contact process, which ultimately produces sulphuric acid. Sulphur dioxide and oxygen are mixed and passed through concentrated sulphuric acid at a controlled rate (see figure 33). The gases next pass over a heated catalyst of either platinum or vanadium pentoxide. The platinum is usually in the form of platinised asbestos, the asbestos increasing the surface area of the platinum. Sulphur trioxide can be collected as a heavy gas or by cooling in ice as a white solid.
   $2SO_2 + O_2 = 2SO_3$

   Industrially, sulphur trioxide is produced in much the same way. Sulphur dioxide, produced by roasting sulphide ores and other methods, is purified, mixed with oxygen (produced from

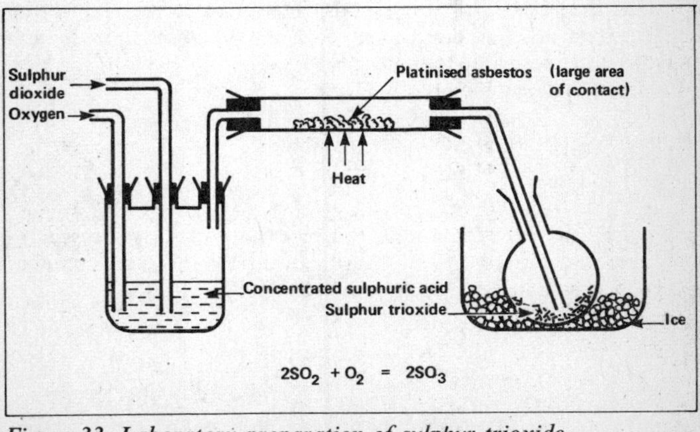

Figure 33. Laboratory preparation of sulphur trioxide

air) and the gas mixture dried. The gases are then passed over a heated catalyst of either platinum or vanadium pentoxide at a temperature of about 450°C. Platinum is easily 'poisoned' and vanadium pentoxide is preferred because it is (a) cheaper and (b) less readily poisoned. The gas formed (sulphur trioxide) is cooled and passes to the absorption tower where it is dissolved in concentrated sulphuric acid which becomes increasingly concentrated and forms a fuming liquid called **oleum.** The oleum is carefully diluted with the correct amount of water to produce ordinary concentrated sulphuric acid.

2. Sulphur trioxide can also be prepared by heating anhydrous iron (II) sulphate ($FeSO_4$).
   $2FeSO_4 = Fe_2O_3 + SO_2\uparrow + SO_3\uparrow$

Properties of sulphur trioxide
It is a white solid consisting of needle-shaped crystals. With water it forms sulphuric acid (it is an acid anhydride).

$$SO_3 + H_2O = H_2SO_4$$

Sulphuric acid
Preparation

1. By the contact process: see sulphur trioxide (figure 33).

2. Heat iron (II) sulphate crystals. The water of crystallisation given off is first condensed and the sulphur dioxide and sulphur trioxide are then dissolved.
   $FeSO_4 7H_2O = FeSO_4 + 7H_2O$
   $2FeSO_4 = Fe_2O_3 + SO_2\uparrow + SO_3\uparrow$
   $SO_2 + H_2O = H_2SO_3$ (unstable)
   $SO_3 + H_2O = H_2SO_4$ (stable)

3. Before the introduction of the modern contact process, sulphuric acid used to be manufactured by the lead chamber process, which may be summarised as follows:

$$\left.\begin{array}{l} 2NO + O_2 = 2NO_2 \\ H_2O + SO_2 = H_2SO_3 \\ H_2SO_3 + NO_2 = H_2SO_4 + NO\uparrow \text{ etc.} \\ 2NO + O_2 = 2NO_2 \end{array}\right\}\text{lead chamber process}$$

Properties of sulphuric acid

It behaves in **four** ways:

1. as a dibasic acid (all strengths);
2. as a drying and dehydrating agent (when concentrated);
3. as an oxidising agent (when hot and concentrated);
4. as a sulphate (all strengths).

Sulphuric acid as an acid

1. It turns blue litmus red.

2. It produces **two** series of salts because it is dibasic, i.e. contains two replaceable hydrogens. If one hydrogen is replaced by a metal, the acid salt (hydrogensulphate) is formed, while if both hydrogens are replaced, the normal salt (sulphate) is formed.
   $NaOH + H_2SO_4 = NaHSO_4 + H_2O$ (acid salt)
   $2NaOH + H_2SO_4 = Na_2SO_4 + 2H_2O$ (normal salt)
   $CuO + H_2SO_4 = CuSO_4 + H_2O$
   (For the experimental details of the above reactions, see page 87.)

3. It will convert sulphates into hydrogensulphates.
   $Na_2SO_4 + H_2SO_4 = 2NaHSO_4$

4. The acid (dilute) produces carbon dioxide with carbonates and hydrogencarbonates, sulphur dioxide with sulphites and hydrogensulphites and hydrogen sulphide with sulphides, etc.
   $Na_2CO_3 + H_2SO_4 = Na_2SO_4 + CO_2\uparrow + H_2O\ (H_2CO_3)$
   $Na_2SO_3 + H_2SO_4 = Na_2SO_4 + SO_2\uparrow + H_2O\ (H_2SO_3)$
   $FeS + H_2SO_4 = FeSO_4 + H_2S\uparrow$
   Concentrated sulphuric acid is a clear oily liquid and is non-volatile. When reacted with the salts of other acids it will displace the acids from their respective salts. Above are typical examples in which weak acids such as carbonic and sulphurous acids are displaced from carbonates and sulphites respectively. Concentrated sulphuric acid will, with chlorides and nitrates, produce the corresponding acids, i.e. hydrochloric and nitric acids respectively. Both are strong acids but are volatile and the standard preparation of each involves the action of a non-volatile acid on the appropriate salt.

$$\left.\begin{array}{l} NaCl + H_2SO_4 = NaHSO_4 + HCl\uparrow \\ NaNO_3 + H_2SO_4 = NaHSO_4 + HNO_3\uparrow \end{array}\right\}\text{concentrated sulphuric acid}$$

5. Dilute sulphuric acid produces hydrogen with zinc, iron magnesium (Z.I.M.) metals.
$Zn + H_2SO_4 = ZnSO_4 + H_2\uparrow$

Sulphuric acid as a drying and dehydrating agent

Sulphuric acid can only act in this way when it is concentrated. The acid has a great affinity for water, and it is as well to note that when diluting the acid with water **the acid should always be added to the water**. Great heat is generated in the process.

1. The acid is hygroscopic, i.e. it absorbs moisture from the air.

2. Concentrated sulphuric acid will remove the elements of water from blue copper (II) sulphate crystals rendering them anhydrous, i.e. the crystals will turn white.
$CuSO_4 5H_2O = CuSO_4 + 5H_2O$
blue white

3. It acts as a drying agent to most gases (N.B. not ammonia, nitrogen monoxide (NO), nitrogen dioxide or hydrogen sulphide).

4. When it is warmed with cane sugar the sugar becomes charred because the elements of water have been removed by the acid.
$C_{12}H_{22}O_{11} = 12C + 11H_2O$
A similar reaction takes place with paper and other hydrocarbons.

5. It removes the elements of water from formic acid and oxalic acid.
$HCOOH = H_2O + CO\uparrow$
$(COOH)_2 = H_2O + CO\uparrow + CO_2\uparrow$

Sulphuric acid as an oxidising agent

When used in this capacity the acid must be **hot and concentrated**. It splits up as follows: $H_2SO_4 - H_2O, SO_2, O$. If sulphuric acid acts as an oxidising agent, it is reduced to sulphur dioxide.

1. It oxidises carbon to carbon dioxide:
$C + 2H_2SO_4 = CO_2 + 2H_2O + 2SO_2$
$2H_2O$
$2SO_2$
$2O$

2. It oxidises sulphur to sulphur dioxide:

$$S + 2H_2SO_4 = 3SO_2 + 2H_2O$$

$2H_2O$
$2SO_2$
$2O$

3. It oxidises hydrogen sulphide to sulphur:

$$H_2S + H_2SO_4 = 2H_2O + SO_2 + S$$

H_2O
SO_2
O

4. It oxidises metals to oxides, then forms sulphates:

$$Cu + H_2SO_4 = CuO + H_2O + SO_2$$

H_2O
SO_2
O

then $CuO + H_2SO_4 = CuSO_4 + H_2O$
adding $Cu + 2H_2SO_4 = CuSO_4 + 2H_2O + SO_2\uparrow$

The residue is grey (white anhydrous copper (II) sulphate and brown copper (I) oxide, Cu_2O). To obtain copper (II) sulphate crystals pour the residue into water, filter and crystallise.

Sulphuric acid as a sulphate

1. With barium chloride ($BaCl_2$) solution a white precipitate of barium sulphate is formed which is insoluble in dilute hydrochloric acid. This is a test for the sulphate ion.
   $H_2SO_4 + BaCl_2 = BaSO_4\downarrow + 2HCl$
2. On heating strongly sulphuric acid breaks down thus:
   $H_2SO_4 = H_2O + SO_2 + \text{'O'}$
   I.e. $2H_2SO_4 = 2H_2O + 2SO_2 + O_2$

Uses of sulphuric acid

1. In the making of fertilisers, e.g. ammonium sulphate, calcium superphosphate.
2. In the manufacture of artificial fibres.
3. In the manufacture of pigments for paints.
4. In the production of hydrochloric acid (salt gas) from salt.
5. Formerly, in the steel industry for cleaning the steel (pickling); now generally superseded in this usage by hydrochloric acid.

Note Sulphuric acid is probably the single most important chemical compound on earth.

Nitrogen (N_2)

Preparation

1. From the air by removing all the other gases (except the inert gases). Air is passed over heated copper to remove oxygen, through potassium hydroxide (or sodium hydroxide) solution to remove carbon dioxide, and through concentrated sulphuric acid to remove water vapour. This nitrogen contains inert gases, e.g. argon, helium, neon, and is often referred to as atmospheric nitrogen.

2. By heating ammonium nitrite. The ammonium nitrite is made in the flask by mixing together sodium nitrite and ammonium chloride because as such ammonium nitrite is unstable. The mixture is heated and nitrogen is evolved.
   $NaNO_2 + NH_4Cl = NaCl + NH_4NO_2$
   $NH_4NO_2 = N_2\uparrow + 2H_2O$

3. By passing ammonia gas over the hot oxide of one of the L.I.C. (lead, iron, copper) metals.
   $2NH_3 + 3CuO = 3Cu + 3H_2O + N_2\uparrow$
   The ammonia is oxidised to nitrogen and the copper (II) oxide reduced to copper.

Properties

Nitrogen is a relatively inert gas. On heating with magnesium the gas combines to form a pale yellow solid, magnesium nitride.

$$3Mg + N_2 = Mg_3N_2$$

Uses of nitrogen

In the manufacture of ammonia and nitric acid.

The nitrogen cycle

Compounds known as proteins contain nitrogen and are essential to all plant and animal life. Plants obtain their protein from the soil while animals obtain proteins by eating plants.

Replenishment of the nitrogen in the soil is achieved:
1. by addition of nitrogenous fertilisers;
2. by decay of plants and animals;
3. by certain plants, e.g. beans, peas, etc. (i.e. leguminous plants).

Thus a cycle is set up.

Ammonia (NH_3)
(the only common alkaline gas)

Preparation

1. Nitrogen from the air, and hydrogen from water (or oil) combine at high pressure (200 atmospheres) and comparatively low temperature (500°C) in the presence of a catalyst (finely divided iron) to form ammonia. This is the well-known Haber process for the industrial preparation of the gas, and a good example of the synthesis of a compound.
   $3H_2 + N_2 \rightleftharpoons 2NH_3$
   Under these conditions 10 per cent of the total volume of gas is converted into ammonia.

2. Ammonia is prepared in the laboratory by heating any ammonium salt with an alkali – usually ammonium chloride and calcium hydroxide respectively.

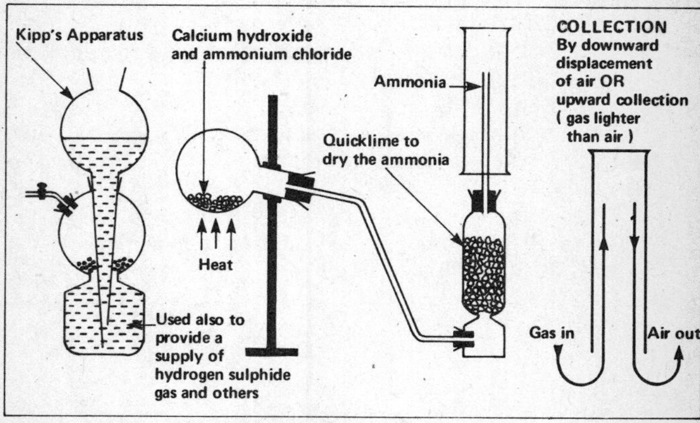

Figure 34. Preparation of ammonia

The gas is collected by the downward displacement of air, and dried by passing up a tower containing quicklime.

$$2NH_4Cl + Ca(OH)_2 = 2NH_3\uparrow + CaCl_2 + 2H_2O$$

Properties of ammonia

1. It is a colourless gas with a pungent smell and taste, turning red litmus blue.
2. It is lighter than air, therefore it can be collected by the downward displacement of air.
3. It is easily liquefied and is used in refrigeration.

4. It is very soluble in water, producing a true base. It is a basic **anhydride**.
   $NH_3 + H_2O = NH_4OH$ (base and alkali)
5. When reacted with acids it gives rise to salts, **but not water**, being a basic anhydride.
   E.g.
   $NH_3 + H_2SO_4 = NH_4HSO_4$ (excess acid)
   $2NH_3 + H_2SO_4 = (NH_4)_2SO_4$ (excess base)
6. It forms dense white fumes of ammonium chloride with hydrogen chloride, when the vapours are allowed to react.
   $NH_3 + HCl = NH_4Cl$ (dense white fumes)
   This is a common test for both gases. If the presence of ammonia is suspected, the stopper of the hydrochloric-acid bottle should be brought near to the gas, and if the presence of hydrogen chloride gas is suspected, the ammonia bottle-stopper is used.
7. It does not burn in air, nor does it support combustion, but it does burn in oxygen.
8. It is oxidised to nitrogen monoxide by passing the ammonia and oxygen over a heated platinum catalyst.
   $4NH_3 + 5O_2 = 4NO\uparrow + 6H_2O$
   This is part of the process leading to the manufacture of nitric acid. The reaction is called the **catalytic oxidation of ammonia.**
9. Ammonia can be oxidised to nitrogen by being passed over a hot oxide of one of the L.I.C. (lead, iron, copper) metals.
   $2NH_3 + 3CuO = 3Cu + 3H_2O + N_2\uparrow$

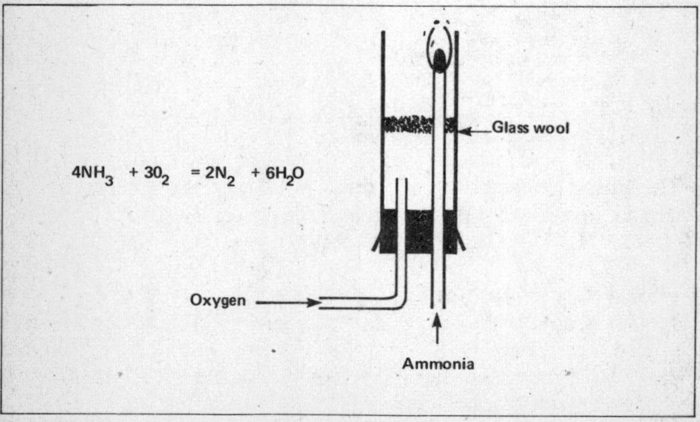

Figure 35. Burning ammonia in oxygen

10. When reacted with chlorine ammonia is oxidised to nitrogen.

$$2NH_3 + 3Cl_2 = N_2\uparrow + 6HCl$$

further $6NH_3 + 6HCl = 6NH_4Cl$

adding $8NH_3 + 3Cl_2 = N_2\uparrow + 6NH_4Cl$

11. Ammonium hydroxide is an important alkali, and is used to precipitate insoluble hydroxides from aqueous solutions of their salts. It produces a brown precipitate of iron (III) hydroxide with iron (III) salts, and a greenish-white precipitate of iron (II) hydroxide with a solution of an iron (II) salt.

    E.g. $FeCl_3 + 3NH_4OH = Fe(OH)_3\downarrow + 3NH_4Cl$

    $FeCl_2 + 2NH_4OH = Fe(OH)_2\downarrow + 2NH_4Cl$

12. Ammonium hydroxide solution gives a light blue precipitate of copper (II) hydroxide with a solution of copper salt.

    $CuSO_4 + 2NH_4OH = Cu(OH)_2\downarrow + (NH_4)_2SO_4$

    The precipitate dissolves in excess producing a deep blue solution of the complex copper ion called tetraammine copper (II) ion.

Tests for ammonia

1. It produces dense white fumes in the presence of hydrochloric acid vapour (HCl stopper test).
   $NH_3 + HCl = NH_4Cl$
2. With copper (II) sulphate it gives a bluish-white precipitate, then a deep blue solution (see 12 above).
3. It turns red litmus blue, being an alkaline gas. As stated above, ammonia is the only common alkaline gas.
4. It has a characteristic smell.

Uses of ammonia

1. In refrigeration.
2. In the manufacture of nitric acid.
3. In the manufacture of the important fertiliser ammonium sulphate.
4. In the manufacture of plastics and resins.

The ammonium ion (NH_4^+)

Ammonium salts are very similar to those of potassium and sodium, e.g. ammonium chloride is a white crystalline solid, soluble in water and on visual examination almost indistinguishable from the chlorides of potassium and sodium. Consequently, there is a similarity between the ions of potassium and sodium, and those of ammonium.

The ammonium ion (see figure 36) is a radical, made up of nitrogen, and four hydrogens, i.e. NH_4^+. The ion is designated NH_4^+, hence the formula for ammonium chloride is NH_4Cl, which indicates that an electron has been donated and accepted.

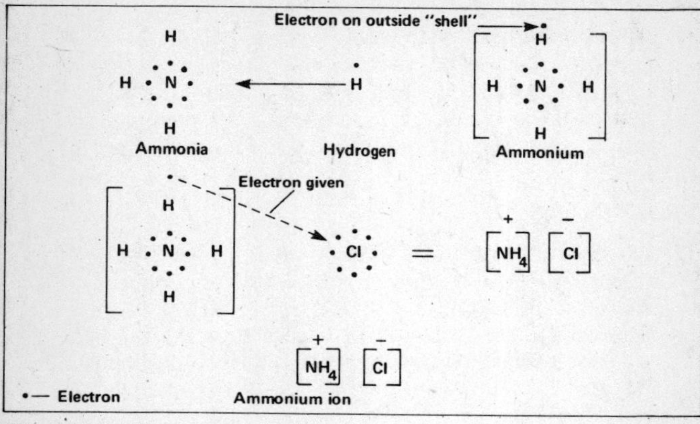

Figure 36. The ammonium ion

The covalent ammonia gas (NH_3) with its lone pair of electrons 'adds on' another hydrogen atom, resulting in the [NH_4] radical having an excess electron, producing an electronic arrangement similar to that of sodium and potassium, i.e. one electron on the outside shell, and similar salts result (NH_4^+, Na^+, K^+).

Ammonium salts

Properties

1. All ammonium salts are soluble in water.
2. All ammonium salts on heating with any base produce ammonia.

$$NH_4Cl + NaOH = NH_4OH + NaCl$$
$$\downarrow$$
$$NH_3 + H_2O$$

3. On heating alone:
   (i) ammonium chloride sublimes and dissociates;
   $NH_4Cl \rightleftharpoons NH_3 + HCl$
   (ii) ammonium nitrate melts and dinitrogen monoxide (N_2O) is evolved;
   (iii) ammonium nitrite gives off nitrogen when heated.
   $NH_4NO_2 = N_2\uparrow + 2H_2O$

Nitric acid (HNO_3)

Preparation

In industry a mixture of ammonia (manufactured by the Haber process) and air is passed over a heated platinum catalyst at a temperature of 700–800°C. The reaction is exothermic and the heat evolved is sufficient to maintain the reaction temperature.

$$4NH_3 + 5O_2 = 4NO + 6H_2O$$

The issuing gases are cooled to about 150°C and mixed with more air when nitrogen dioxide is produced.

$$2NO + O_2 \rightleftharpoons 2NO_2$$

The nitrogen dioxide and excess air is passed up a water-cooled steel tower down which water is flowing, and nitric acid is produced.

$$2NO_2 + H_2O = HNO_3 + HNO_2 \text{ (nitrous acid)}$$

The nitrous acid formed decomposes, liberating nitrogen monoxide which is oxidised by the air to nitrogen dioxide; this then reacts with water.

$$3HNO_2 = HNO_3 + 2NO + H_2O$$

In the laboratory: by heating any nitrate with the less volatile concentrated sulphuric acid. Sodium or potassium nitrate are most commonly used. The nitrate and acid are warmed gently in a retort (figure 37).

$$NaNO_3 + H_2SO_4 = NaHSO_4 + HNO_3\uparrow \text{ (gentle heat)}$$
$$NaNO_3 + NaHSO_4 = Na_2SO_4 + HNO_3\uparrow \text{ (strong heat)}$$

Compare the above reactions with those for the preparation of hydrogen chloride.

$$NaCl + H_2SO_4 = NaHSO_4 + HCl\uparrow$$
$$NaCl + NaHSO_4 = Na_2SO_4 + HCl\uparrow$$

Precautions in the preparation of nitric acid

1. The apparatus must be entirely of glass because the acid, being a very strong oxidising agent, will attack such materials as cork or rubber.
2. Only gentle heat should be applied, to minimise the splitting up of the nitric acid when formed.
3. Gentle heat should be used, in order to distil the nitric acid and not the sulphuric acid.

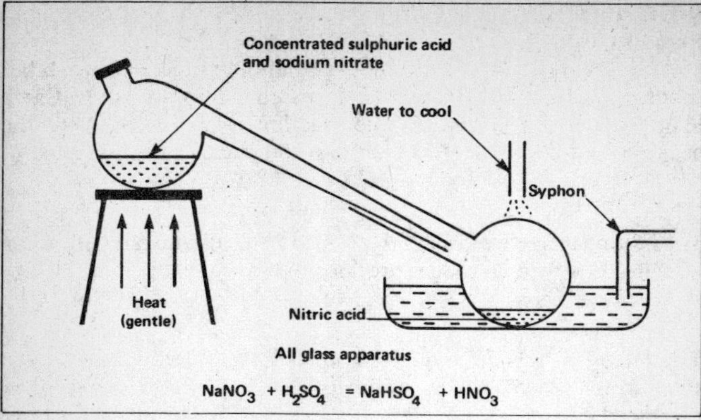

Figure 37. Preparation of nitric acid

Properties of nitric acid

It can act in three ways:

1. as an acid (at all strengths);
2. as an oxidising agent (best when concentrated);
3. as an oxidising agent and an acid (at all strengths).

Nitric acid as an acid

1. It turns blue litmus red.

2. **At all strengths** it gives salts with bases.
   E.g. $CuO + 2HNO_3 = Cu(NO_3)_2 + H_2O$

3. **At all strengths** it will liberate carbon dioxide from carbonates and hydrogencarbonates.
   E.g. $CaCO_3 + 2HNO_3 = Ca(NO_3)_2 + H_2CO_3\ (H_2O + CO_2)$
   $NaHCO_3 + HNO_3 = NaNO_3 + H_2CO_3\ (H_2O + CO_2)$

4. **Very dilute** nitric acid reacts with magnesium, liberating hydrogen.
   $Mg + 2HNO_3 = Mg(NO_3)_2 + H_2\uparrow$

Nitric acid as an oxidising agent

When acting in this way the acid may be regarded as splitting up thus:

$$2HNO_3 \text{ —— } H_2O,\ 2NO_2,\ O$$

Note When nitric acid **oxidises**, it is itself reduced.

1. **Hot concentrated** nitric acid oxidises carbon to carbon dioxide. (It causes hot charcoal or sawdust to burst into flame.)

$C + 4HNO_3 = CO_2 + 2H_2O + 4NO_2$
$\quad 2H_2O$
$\quad 4NO_2$
$\quad 2O$

2. **Hot concentrated** nitric acid oxidises sulphur to sulphur trioxide, which forms sulphuric acid in solution.

$S + 6HNO_3 = SO_3 + 3H_2O + 6NO_2$
$\quad 3H_2O \qquad (H_2SO_4 + H_2O)$
$\quad 6NO_2$
$\quad 3O$

3. **Cold concentrated** nitric acid oxidises hydrogen sulphide to sulphur (yellow precipitate); if **hot** the sulphur would be oxidised to sulphuric acid.

$H_2S + 2HNO_3 = S + 2H_2O + 2NO_2$
$\quad H_2O$
$\quad 2NO_2$
$\quad O$

4. **Cold concentrated** nitric acid oxidises sulphur dioxide and sulphites to sulphates.

$H_2SO_3 + 2HNO_3 = H_2SO_4 + H_2O + 2NO_2$
$\quad H_2O$
$\quad 2NO_2$
$\quad O$

Nitric acid acting as an oxidising agent and an acid

1. **Cold diluted** nitric acid and the common metals liberate nitrogen monoxide and the corresponding salt is formed.
$3Cu + 8HNO_3 = 3Cu(NO_3)_2 + 2NO\uparrow + 4H_2O$

2. **Hot concentrated** nitric acid with the common metals will produce the corresponding salt and nitrogen dioxide.
$Cu + 2HNO_3 = CuO + H_2O + 2NO_2\uparrow$ (oxidation of copper and reduction of nitric acid)
$CuO + 2HNO_3 = Cu(NO_3)_2 + H_2O$ (acid and base)
$Cu + 4HNO_3 = Cu(NO_3)_2 + 2H_2O + 2NO_2\uparrow$

3. Concentrated nitric acid renders iron passive, because of the formation of an oxide layer on the surface of the iron.

Tests for nitric acid

1. It turns blue litmus red.

2. It will produce a **brown ring** compound with iron (II) sulphate solution and concentrated sulphuric acid (see tests for nitrates, below).

3. Nitric acid (strong) produces nitrogen dioxide with copper.

Uses of nitric acid

1. In the manufacture of fertiliser, e.g. sodium nitrate.
2. In the manufacture of explosives, e.g. TNT (trinitrotoluene).
3. In the preparation of substances from which dyes are made.

Nitrates

Properties

1. All nitrates are soluble in water.

2. All nitrates on being warmed with concentrated sulphuric acid and copper give off the brown gas nitrogen dioxide (NO_2). This is a test for nitrates.
$$NaNO_3 + H_2SO_4 = NaHSO_4 + HNO_3$$
$$Cu + 4HNO_3 = Cu(NO_3)_2 + 2H_2O + 2NO_2\uparrow$$

3. All nitrates give a white precipitate with **nitron** when in solution – yet another test for nitrates.

4. All nitrates can be detected by the brown ring test. Freshly prepared iron (II) sulphate solution is added to a solution of the nitrate. Concentrated sulphuric acid is poured slowly down the side of the tube, and a brown ring appears at the junction of the liquids. The test can be explained thus:
   (i) firstly nitric acid is liberated:
   $$NaNO_3 + H_2SO_4 = NaHSO_4 + HNO_3$$
   (ii) then nitrogen monoxide is formed:
   $$6FeSO_4 + 3H_2SO_4 + 2HNO_3 = 3Fe_2(SO_4)_3 + 2NO + 4H_2O$$
   (iii) nitrogen monoxide (NO) then forms the brown ring:
   $$FeSO_4 + NO = FeSO_4NO$$
   nitroso-iron (II) sulphate

Action of heat on nitrates

1. Sodium and potassium nitrates melt and turn yellow, oxygen is evolved and the nitrite remains.
   $2NaNO_3 = 2NaNO_2 + O_2\uparrow$

2. Ammonium nitrate decomposes to produce dinitrogen monoxide and steam.
   $NH_4NO_3 = N_2O\uparrow + 2H_2O$

3. All other nitrates, i.e. the nitrates of 'heavy metals', produce nitrogen dioxide and oxygen, leaving the oxide as the residue.

   $2Pb(NO_3)_2 = 2PbO + 4NO_2\uparrow + O_2\uparrow$

Note If the oxide is unstable, then the metal and oxygen are produced.
E.g. $2AgNO_3 = 2Ag + 2NO_2\uparrow + O_2\uparrow$

Chlorine (Cl_2)

Preparation

Industrially the gas is prepared as a by-product in the manufacture of sodium hydroxide, and the electrolysis of fused sodium chloride.

In the laboratory any of the following methods may be used.

1. Oxidise concentrated hydrochloric acid with any oxidising agents.
   'O' (oxidising agent) $+ 2HCl = H_2O + Cl_2$ (key equation)

   e.g. $MnO_2 + 4HCl = 2H_2O + MnCl_2 + Cl_2\uparrow$ (heat)

   $PbO_2 + 4HCl = 2H_2O + PbCl_2 + Cl_2\uparrow$ (heat)

   $Pb_3O_4 + 8HCl = 4H_2O + 3PbCl_2 + Cl_2\uparrow$ (heat)

   Chlorine can be prepared from potassium permanganate and concentrated hydrochloric acid **without heat.**

2. Treat any chloride with manganese (IV) oxide and concentrated sulphuric acid.

E.g. $NaCl + H_2SO_4 = NaHSO_4 + HCl$

$MnO_2 + 4HCl = 2H_2O + MnCl_2 + Cl_2\uparrow$

3. Treat bleaching powder with **any** acid.
E.g. $CaOCl_2 + H_2SO_4 = CaSO_4 + H_2O + Cl_2\uparrow$

If the gas is required pure and dry it is bubbled through water to remove any hydrogen chloride, through concentrated sulphuric acid to dry the gas, then collected by the **upward displacement of air** (the gas is heavier than air).

Properties of chlorine

1. It is a greenish-yellow, poisonous gas with a sharp taste and suffocating smell.

2. It is much heavier than air.

3. It dissolves in water producing two acids, hydrochloric acid and hypochlorous acid.
$Cl_2 + H_2O = HCl + HClO$
↓ hydrochloric acid (HCl) ↓ hypochlorous acid (HClO)

4. It does not burn, but supports the combustion of some substances (see 5(i) below).

5. It has a very strong affinity for hydrogen and is therefore a very strong **oxidising agent**.

   (i) It combines directly with hydrogen in light (not in the dark). A mixture of hydrogen and chlorine explodes in sunlight. A hydrogen flame continues to burn if placed in chlorine.
   $H_2 + Cl_2 = 2HCl$

   (ii) It combines readily with hydrogen in hydrocarbons, e.g. turpentine burns with a smoky flame in chlorine.
   $C_{10}H_{16} + 8Cl_2 = 10C + 16HCl$

(iii) It combines readily with hydrogen in ammonia.

$$2NH_3 + 3Cl_2 = N_2 + \cancel{6HCl} \text{ ——————(1)}$$

$$6NH_3 + \cancel{6HCl} = 6NH_4Cl \text{ ——————(2)}$$

$$8NH_3 + 3Cl_2 = NH_4Cl + N_2 \text{ adding (1) \& (2)}$$

(iv) It combines readily with hydrogen in hydrogen sulphide.
$H_2S + Cl_2 = 2HCl + S$ (yellow cloud)

6. It combines readily with some metals.
E.g. $2Na + Cl_2 = 2NaCl$ (sodium burns)

$Mg + Cl_2 = MgCl_2$ (magnesium burns)

$2Fe + 3Cl_2 = 2FeCl_3$

In the reaction with iron, dry chlorine is passed over **heated** iron.

7. Yellow phosphorus ignites in chlorine.
$2P + 3Cl_2 = 2PCl_3$ (limited chlorine)

$2P + 5Cl_2 = 2PCl_5$ (excess chlorine)

8. With water it forms **chlorine water** which contains hypochlorous acid (HClO). This acid is unstable and breaks down thus:
$HClO = HCl + $ 'O'
nascent oxygen

This forms molecular oxygen when exposed to sunlight and the greenish colour of the chlorine water disappears. The oxygen released by the breakdown of the unstable hypochlorous acid bleaches vegetable matter by oxidation.

9. It displaces bromine from bromides and iodine from iodides.
$2KBr + Cl_2 = 2KCl + Br_2$

$2KI + Cl_2 = 2KCl + I_2$

Tests for chlorine

1. It is greenish-yellow in colour and has a characteristic smell.

2. It bleaches moist litmus paper after turning it red.

3. It displaces iodine from potassium iodide solution.
$2KI + Cl_2 = 2KCl + I_2$

Starch iodide paper is used in this test, the iodine produced turning the starch blue.

Uses of chlorine
1. As a bleach.
2. As a germicide and disinfectant.
3. In the manufacture of tetrachloromethane (carbon tetrachloride) which is used in fire extinguishers.

Note Chlorine also reacts readily with alkalis.

(i) **Cold dilute solution** Chlorine reacts to form the chloride, the hypochlorite and water.
E.g. $Cl_2 + 2NaOH = NaCl + NaClO + H_2O$
($NaClO$: sodium hypochlorite)

(ii) **Hot concentrated solution** A mixture of the chloride, chlorate and water is formed.
$6NaOH + 3Cl_2 = 5NaCl + NaClO_3 + 3H_2O$
($NaClO_3$: sodium chlorate)

Hydrogen chloride (hydrochloric acid) (HCl)

Commonly called **salt gas** because it can be prepared from common salt (sodium chloride).

Preparation
1. Burn hydrogen in chlorine.
$H_2 + Cl_2 = 2HCl$
(Yet another example of the **synthesis** of a compound.)
2. Warm a chloride with concentrated sulphuric acid.
E.g. $NaCl + H_2SO_4 = NaHSO_4 + HCl\uparrow$ gentle heat
$2NaCl + H_2SO_4 = Na_2SO_4 + 2HCl\uparrow$ strong heat
Compare the above reactions with:
$NaNO_3 + H_2SO_4 = NaHSO_4 + HNO_3\uparrow$ gentle heat
$2NaNO_3 + H_2SO_4 = Na_2SO_4 + 2HNO_3\uparrow$ strong heat

The preparation is carried out in the laboratory by warming a mixture of rock salt and sulphuric acid in a flask, thistle funnel, and delivery tube apparatus. (Rock salt is impure sodium chloride.)

The gas is collected by the upward displacement of air and dried by passing through concentrated sulphuric acid.

Properties of hydrogen chloride

1. A colourless gas with a sharp taste and smell.

2. It fumes in moist air, and is heavier than air.

3. It is very soluble in water, giving an acid solution. To dissolve the gas in water an 'anti-suck-back' device is used (see figure 38).

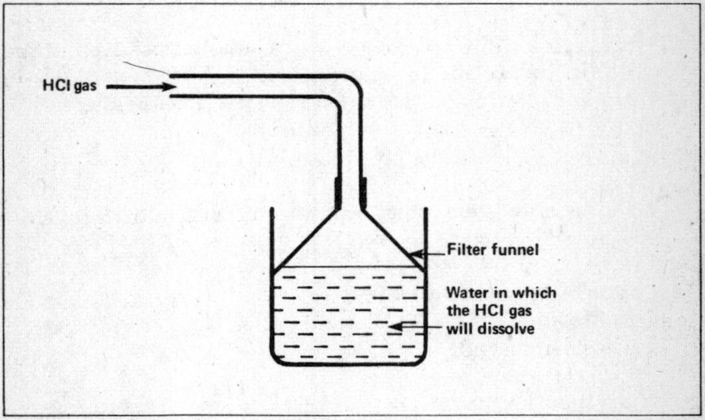

Figure 38. Anti-suck-back device

4. It does not burn or support combustion.

5. It produces salts when reacted with bases. Only one series of salts is possible because it is a monobasic acid, i.e. it only has one replaceable hydrogen atom.
   E.g. $ZnO + 2HCl = ZnCl_2 + H_2O$
   base + acid = salt + water

6. It liberates carbon dioxide from carbonates and hydrogen-carbonates, sulphur dioxide from sulphites and hydrogen-sulphites.

E.g. $Na_2CO_3 + 2HCl = 2NaCl + H_2CO_3\ (H_2O + CO_2)$
$NaHCO_3 + HCl = NaCl + H_2CO_3\ (H_2O + CO_2)$
$Na_2SO_3 + 2HCl = 2NaCl + H_2SO_3\ (H_2O + SO_2)$
$NaHSO_3 + HCl = NaCl + H_2SO_3\ (H_2O + SO_2)$

7. When reacted with the Z.I.M. (zinc, iron, magnesium) metals, hydrogen is liberated.
E.g. $Zn + 2HCl = ZnCl_2 + H_2\uparrow$

8. Hydrochloric acid produces chlorine when reacted upon by oxidising agents.
$\text{'O'} + 2HCl = H_2O + Cl_2\uparrow$
oxidising
agent
E.g. $MnO_2 + 4HCl = 2H_2O + MnCl_2 + Cl_2\uparrow$ (heat)

9. It produces a white precipitate with a solution of silver nitrate which is **insoluble** in dilute nitric acid but **soluble** in ammonium hydroxide. **This is a test for a chloride.**
$AgNO_3 + HCl = AgCl\downarrow + HNO_3$
white precipitate

10. It forms dense white fumes with an ammonia-bottle stopper.
$NH_3 + HCl = NH_4Cl$

Tests for hydrogen chloride (hydrochloric acid)

1. It fumes in moist air.

2. It turns blue litmus red.

3. It forms dense white fumes with an ammonia-bottle stopper.

4. In solution it produces a white precipitate with silver nitrate (see 9, above).

Chlorides

Preparation

By the action of hydrochloric acid on (i) a metal; (ii) an oxide or hydroxide; (iii) a carbonate. For details see chapter 8, page 82.

Anhydrous chlorides

To obtain the **higher** chloride the metal should be heated in **chlorine.** To obtain the **lower** chloride the metal should be heated in hydrogen chloride. E.g. to obtain iron (III) chloride (the **higher** chloride):

$$2Fe + 3Cl_2 = 2FeCl_3$$

while to obtain the **lower** chloride, iron (II) chloride

$$Fe + 2HCl = FeCl_2 + H_2\uparrow$$

Properties and tests for chlorides

1. All common chlorides are soluble in water except silver chloride ($AgCl$) and lead (II) chloride ($PbCl_2$).

2. All chlorides on being warmed with concentrated sulphuric acid produce hydrogen chloride gas.
   E.g. $NaCl + H_2SO_4 = NaHSO_4 + HCl\uparrow$

3. All chlorides on being warmed with concentrated sulphuric acid and manganese (IV) oxide produce chlorine.
   E.g. $NaCl + H_2SO_4 = NaHSO_4 + HCl\uparrow$ } HCl generated
   $MnO_2 + 4HCl = 2H_2O + MnCl_2 + Cl_2\uparrow$ } HCl reacts with MnO_2

4. All chlorides in solution produce a white precipitate with silver nitrate solution, which is insoluble in dilute nitric acid but soluble in ammonium hydroxide.
   E.g. $NaCl + AgNO_3 = AgCl\downarrow + NaNO_3$

The halogens

The halogens, being a typical chemical family, are included in the syllabuses of several examining boards. Whereas chlorine, a member of this family, has already been discussed in detail, this section will explain the overall group or family relationship rather than discussing each element as a separate entity.

This family of elements is to be found in Group 7 of the Periodic table, and as such each element has seven electrons on its outside shell.

The members of the group are:

fluorine	2	7				
chlorine	2	8	7			
bromine	2	8	18	7		
iodine	2	8	18	18	7	
astatine	2	8	18	32	18	7

All the members of the group are non-metallic although their characters become increasingly metallic in accordance with how far down the group they appear. Thus:

fluorine is a yellow gas;
chlorine is a greenish-yellow gas;
bromine is a dark red fuming liquid;
iodine is a grey solid;
astatine is a dark grey or black solid.

(Another interesting feature is that of the darkening in colour of the elements as the atomic number increases.)

1. Reactivity of the halogens

The order of reactivity is fluorine, chlorine, bromine, iodine, astatine. This can be easily demonstrated.

E.g. Chlorine will displace bromine from a solution of a bromide.

$$Cl_2 + 2NaBr = 2NaCl + Br_2$$

Bromine will displace iodine from a solution of an iodide.

$$Br_2 + 2NaI = 2NaBr + I_2$$

Iodine will not displace bromine or chlorine from bromides and chlorides respectively.

2. Oxidising properties of the halogens

All the halogens are oxidising agents because they can be reduced, e.g. they all react with hydrogen to produce the corresponding acids:

Acid	Formula	Salt
hydrofluoric acid	HF	fluorides
hydrochloric acid	HCl	chlorides
hydrobromic acid	HBr	bromides
hydroiodic acid	HI	iodides

(The acid formed in the case of astatine is not relevant to this book.)

3. Acids of the halogens

These are listed in (2) above. All the acids are **reducing agents** because they can be **oxidised**.

E.g. 'O' + 2H.halogen = H_2O + halogen (Cl_2, Br_2, I_2, etc.)
oxidising
agent

$MnO_2 + 2HCl = H_2O + Cl_2 + MnCl_2$

However, their reducing nature increases from fluorine to iodine, i.e. hydrofluoric acid is not as good a reducing agent as hydroiodic acid, because hydroiodic can be oxidised much more easily than hydrofluoric acid.

The reducing powers of the acids can be illustrated as follows.

To prepare any acid the corresponding salt of the acid is treated with concentrated sulphuric acid and the acid is displaced from its salt.
E.g. $NaX + H_2SO_4 = NaHSO_4 + HX$ (X = halogen)

It follows that if a chloride is treated with concentrated sulphuric acid, hydrogen chloride would be produced. This is in fact the case. However if bromides or iodides are reacted with concentrated sulphuric acid, bromine and iodine respectively are formed.

The reasons for this are as follows.

$NaBr + H_2SO_4 = NaHSO_4 + HBr$

Hydrogen bromide is produced as illustrated by the above equation. Work already covered on sulphuric acid has shown that it is a strong oxidising agent, therefore as soon as the hydrogen bromide (a strong reducing agent) is produced, the sulphuric acid will immediately oxidise it to bromine.

'O' + $2HBr = H_2O + Br_2$
oxidising agent
↓
sulphuric acid

This reaction also occurs with hydrogen iodide.

$NaI + H_2SO_4 = NaHSO_4 + HI$

'O' + $2HI = H_2O + I_2$
oxidising agent
↓
sulphuric acid

In the case of hydrogen chloride, which is not such a powerful reducing agent as HBr and HI, and therefore not as easily oxidised, hydrogen chloride is formed.

To prepare hydrogen bromide and hydrogen iodide, the respective salts must be reacted with a strong, non-volatile acid which is not a strong oxidising agent. Such an acid is phosphoric acid (H_3PO_4). If sodium bromide were warmed with syrupy phosphoric acid, hydrogen bromide would be formed. A similar reaction would occur in the case of the iodide.

To convert chloride ions into chlorine atoms simply oxidise the chloride ion.

E.g. $2Cl^- - 2e = Cl_2$ (oxidation)

$2HCl + \text{'O'} = H_2O + Cl_2$ (oxidation)

('O' = oxidising agent)

To convert sodium chloride into chlorine, hydrogen chloride is first produced, and then oxidised.

4. Tests for the halogens

(i) **In solution** Add a few drops of silver nitrate solution, followed by excess ammonium hydroxide, to a solution of the halogen:

Chloride A white precipitate of silver chloride is produced which **dissolves readily** in ammonium hydroxide.

Bromide A pale yellow precipitate is formed which is **slightly soluble** in ammonium hydroxide.

Iodide A primrose-yellow precipitate is formed which is **insoluble** in ammonium hydroxide.

(ii) **In the solid state** Add to a small quantity of each solid a few drops of concentrated sulphuric acid:

Chloride Fumes of hydrogen chloride are evolved which will form a white smoke if tested with the ammonia-bottle stopper.

Bromide Dark red droplets of bromine are formed and the reddish-brown bromine vapour evolved.

Iodide A violet coloration of iodine vapour is observed.

Chapter 12
Organic Chemistry

History

Before the nineteenth century it was considered impossible to make organic compounds, e.g. sugars, oils, alcohols, without the influence of a 'vital force', which could only be found in living matter. However, in 1845 acetic acid was synthesised from carbon, hydrogen and oxygen, and in 1928 urea was prepared synthetically. Now organic chemistry remains a convenient subdivision and deals with all carbon compounds except CO_2 and carbonates.

The carbon atom

Carbon is unique among the elements because of the very large number of compounds it can form. It appears in Group 4 of the Periodic table, hence it has four electrons on its outside shell. It is half-way along the period, showing almost equal affinity for the elements to the right and left of it, and because of this it can form compounds having very different properties.

Carbon atoms are able to form stable links with each other, in the form of chains (which can be branched, or joined end-to-end in a ring-like structure). No other element has atoms which exhibit this characteristic to such a degree.

Classification of compounds

Aliphatic compounds are those compounds containing 'straight' chains of carbon atoms.

Aromatic compounds are those compounds which are usually joined end to end in a cyclic fashion, making up a 'ring' of six carbon atoms.

These two groups of compounds can be subdivided into **homologous series**, which may be considered a series of organic compounds in which the molecular formula of one compound differs from the next by an 'increment', CH_2. Each member of the series is called a **homologue.** All members have similar chemical properties and show a steady gradation in physical properties with rise in molecular weight. A good example of a homologous series is the **alkanes**, which have the general formula of C_nH_{2n+2}.

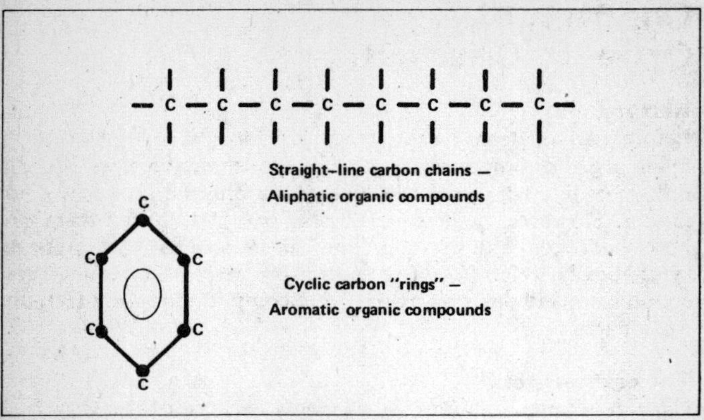

Figure 39. Types of carbon compounds

The alkanes (paraffins)

Name	**Molecular formula**	**Structural formula**
methane	CH_4	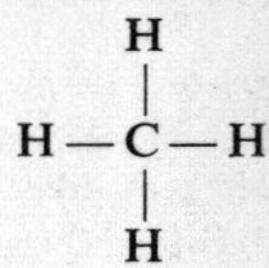
ethane	C_2H_6 CH_3CH_3	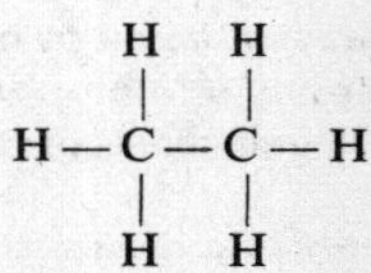
propane	C_3H_8 $CH_3CH_2CH_3$	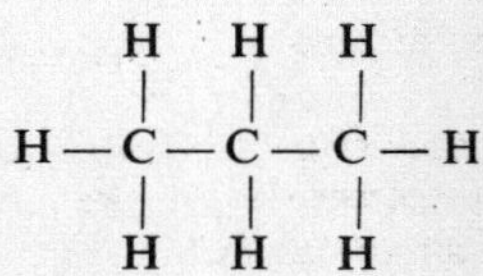

As can be seen the number of compounds possible in this series is enormous, each new member being $-CH_2$ different from its neighbour.

Mode of reaction of the alkanes

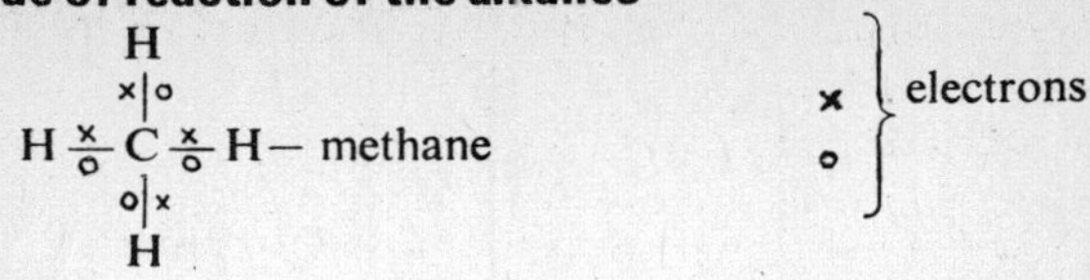

Above is a molecule of methane, the four electrons of carbon sharing with the four electrons from the hydrogens to give the octet of electrons.

If methane reacted, it would have to do so by **substitution** of one or more of the hydrogen atoms; it could not add on any atom or group of atoms because it has no available bonds. The chemistry of methane is therefore one of **substitution reactions**. When an organic chemical reacts by substitution it is said to be **saturated**.

E.g.

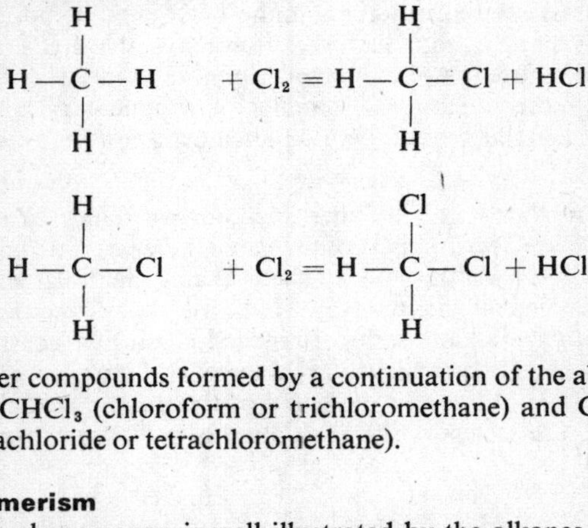

Other compounds formed by a continuation of the above process are $CHCl_3$ (chloroform or trichloromethane) and CCl_4 (carbon tetrachloride or tetrachloromethane).

Isomerism

This phenomenon is well illustrated by the alkanes, particularly those after propane in the series. Compounds having the same molecular formula but different structural formulae are called **isomers**. Butane (C_4H_{10}) is the first of the alkanes to exhibit isomerism, e.g. one structure of butane could be

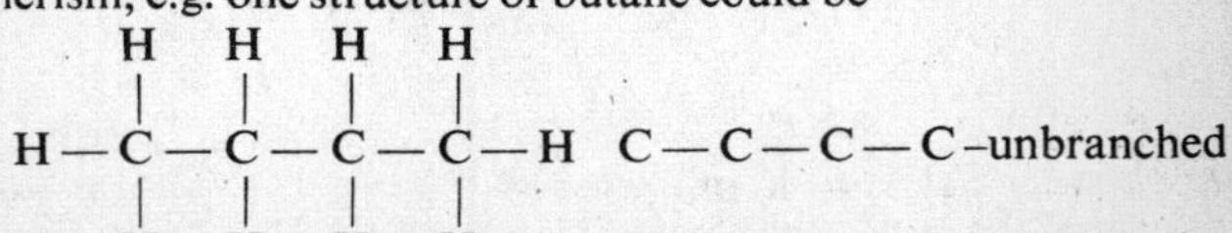

i.e. $CH_3—CH_2—CH_2—CH_3$

However, the carbon atoms could, instead of being in a straight chain, branch thus:

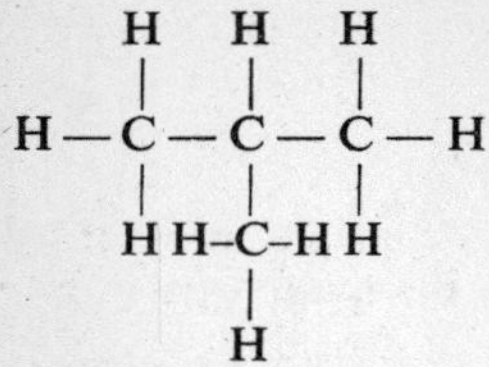

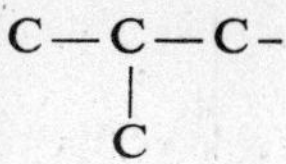
branched

$CH_3—CH(CH_3)—CH_3$

Both compounds would have the same molecular formula C_4H_{10} but the structural formulae are different as illustrated.

Nomenclature

Until comparatively recently the naming of organic compounds was at times haphazard. However, nowadays the I.U.P.A.C. nomenclature is used. Although you do not yet need detailed knowledge of this method of naming, a rudimentary one is useful. Again the alkanes can be used for illustration purposes.

As stated previously the alkanes include **meth**ane, **eth**ane, **prop**ane, **but**ane. The first part of the name indicates the number of carbon atoms in the unbranched chain, meth–1, eth–2, prop–3, but–4. The second part (-ane) indicates that it is saturated and belongs to the alkane series. Branched chain hydrocarbons are named by using a combination of the name of the alkyl group (CH_3, C_2H_5, C_3H_7, etc.) and the name of the unbranched chain hydrocarbon. E.g. consider the compound of molecular formula C_4H_{10}.

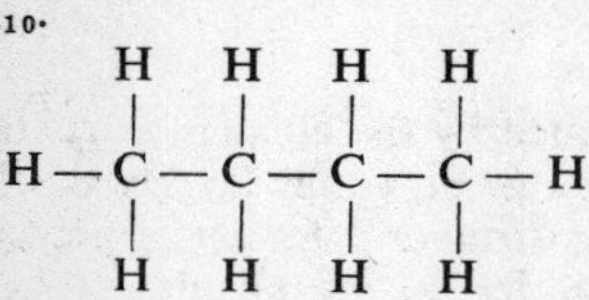

I

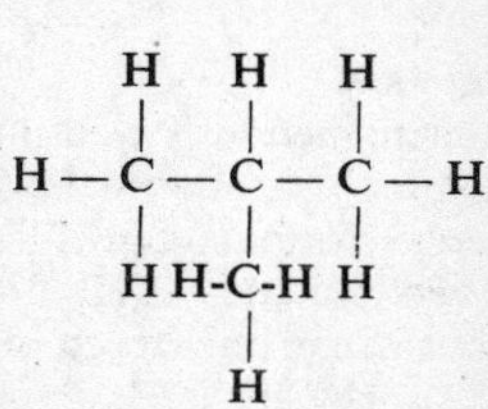

II

Compound I has an unbranched chain of 4 carbon atoms (but-), is saturated and belongs to the alkane series (-ane); its name is **butane.**

Compound II has 3 carbon atoms in the unbranched chain (prop-) but it has 1 carbon atom branching from the second carbon atom in the chain C — C — C.
with a C bonded below the middle C

On this carbon atom are three hydrogens in all comprising the $-CH_3$ or methyl group (one of the alkyl groups). Therefore the compound's name is made up as follows: 2–methyl (i.e. a methyl group on the second carbon atom); propane (3 carbons in the chain of a saturated compound, and one of the alkanes); its name is therefore **2-methyl propane.** Other examples of isomerism are:

C_5H_{12}:

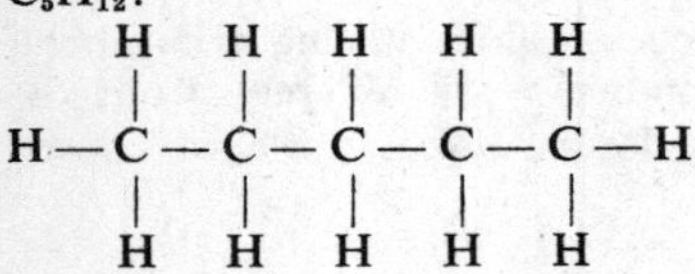

pentane

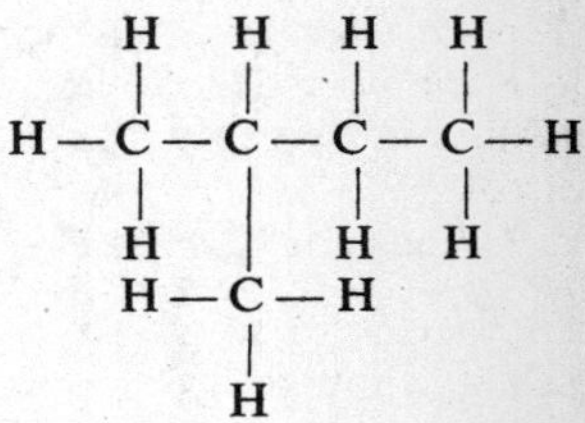

2-methyl butane

C_2H_6O

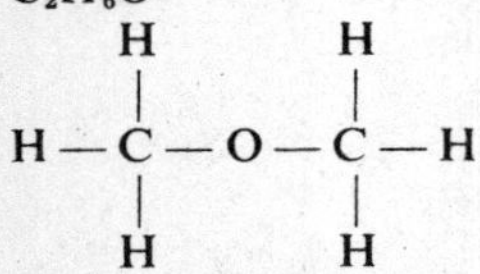

dimethyl ether

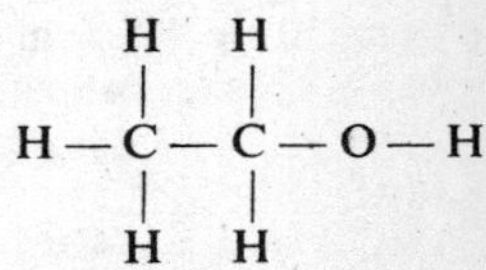

ethanol

The compound C_5H_{12} has another isomer. For practice try to derive its structure and name.

Properties of the alkanes

1. C_1 to C_4 are gases (methane being the main constituent of natural gas) C_5 to C_{17} are liquids, while C_{18} and above are solids.
2. The alkanes are chemically inert because all the atoms are joined by single covalent bonds and are therefore saturated, but they all:
   (i) burn to give carbon monoxide and water;

(ii) are easily chlorinated in diffuse sunlight with the hydrogen being easily substituted.

Uses of alkanes

1. Methane occurs in natural gas and is therefore a fuel.
2. Methane can be broken down into hydrogen and carbon (lampblack), which is used in the manufacture of inks, carbon paper, etc.
3. Propane and butane are well-known gases used mainly as fuels, burning to give carbon dioxide.
4. Alkanes are also used extensively in the manufacture of solvents and anaesthetics.

The alkenes (olefins)

Another example of a homologous series having the general formula of C_nH_{2n}. These compounds are different from the alkanes in that they all contain double bonds (or have unused bonds).

E.g. ethene C_2H_4

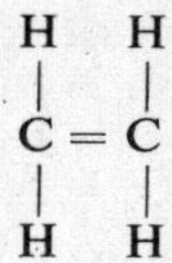

This means that the alkenes react by **addition** and not substitution, and as such are said to be **unsaturated** hydrocarbons ('–ene' in the name indicating unsaturation).

E.g. When ethene reacts it 'adds on' atoms

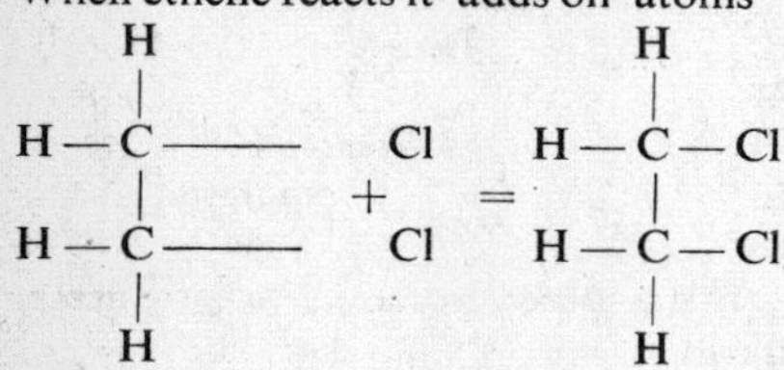

The old theory was that the double bond joining the carbons (C = C) broke and the atoms added on to the broken halves. This idea has now given way to a more modern concept which is part of the A-level course.

Other members of the alkene series are propene (C_3H_6), butene (C_4H_8), etc. It is interesting to note that there is no single carbon compound which would if it existed be called methene.

Polymerisation

A **polymer** is a very large molecule, being made from many (poly) much smaller molecules called monomers. Ethene undergoes polymerisation to polyethene (polythene). The ethene monomers link up with each other to form molecules having molecular weights of about 25,000–30,000.

monomer

$$CH_2 = CH_2 + CH_2 = CH_2 + CH_2 = CH_2 \text{ etc.}$$

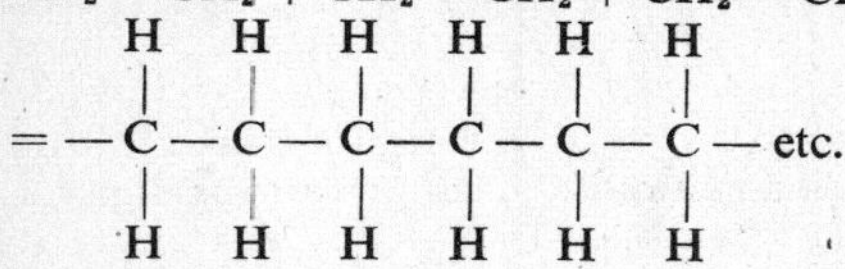

polymer

Nylon and terylene are both examples of polymers.

To distinguish between a saturated hydrocarbon (methane) and an unsaturated hydrocarbon (ethene)

This is a popular question in the O–level examinations. The answer is to pass each gas in turn through bromine. The saturated compound will have no effect on the halogen but the unsaturated compound will decolorise the halogen because of the formation of an addition compound.

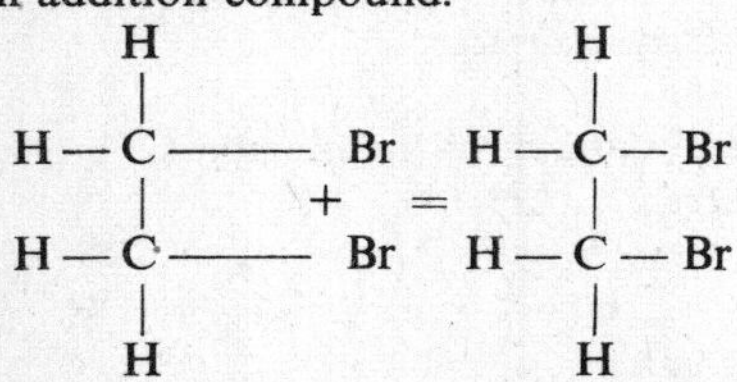

(colourless dibromoethane)

Oil

Until quite recently crude petroleum, or crude oil as it is known, was something which most consumers took for granted, but the huge price increases in this product have aroused much greater interest in it and increased awareness of its importance.

Petroleum is a mixture of paraffins which are separated by fractional distillation (see chapter 2). Some of the products of the distillation process are usable as such; however the majority have to be processed further by means of **catalytic cracking**. This process involves the breaking down of a large molecule into smaller parts, using catalysts.

E.g. $C_{10}H_{22} = C_7H_{16} + C_3H_6$

The quite large 10-carbon molecule has been split (cracked) into a C_7 and a C_3 molecule, from which many common products can be made.

Alcohols

All these compounds contain the $-OH$ or hydroxyl group.

E.g. ethanol

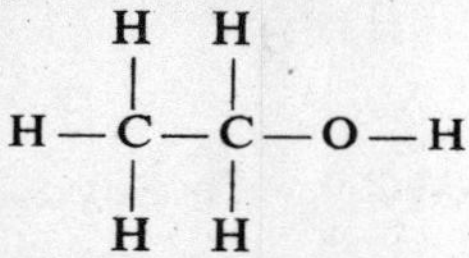

Ethanol is the alcohol commonly found in beers, wines and spirits, and as such is produced by the process of fermentation. It is one of yet another group of compounds which belong to the homologous series, the **alcohols**.

methanol	CH_3OH
ethanol	C_2H_5OH
propanol	C_3H_7OH etc.

Note once again the logical naming of the members.

E.g. propanol C_3H_7OH

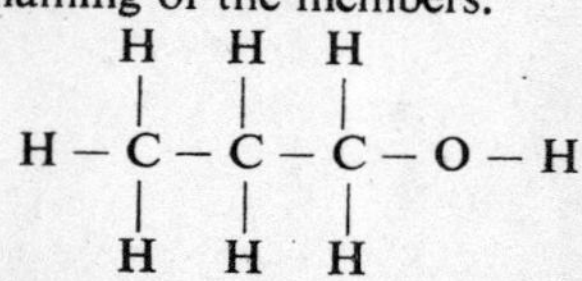

3 carbon atoms in the chain (prop-), saturated (-an), an alcohol (-ol): prop-an-ol.

Ethanol (C_2H_5OH) is typical of the alcohols and is well known. Its properties typify those of the alcohol series.

1. It burns to form water and carbon dioxide.
   $C_2H_5OH + 3O_2 = 2CO_2 + 3H_2O$

2. It can be oxidised, in common with all alcohols, to the corresponding acid.

E.g. $C_2H_5OH + 2(O) = CH_3COOH + H_2O$
oxidising agent acetic acid
or $C_2H_5OH + (O) = CH_3COOH$
Acetic acid is also known as ethanoic acid.

3. In inorganic chemistry a well-known general reaction is:
acid + base = salt + water

An equally well-known reaction in organic chemistry is:

acid	+ alcohol	⇌ ester	+ water
CH_3COOH	$+ C_2H_5OH$	$\rightleftharpoons CH_3COOC_2H_5$	$+ H_2O$
↓	↓	↓	↓
ethanoic acid	+ ethanol	⇌ ethyl acetate	+ water

Ethyl acetate is called an ester. Esters are organic compounds commonly used in the cosmetic industry for perfumes etc. because of their sweet smell. Ethanoic acid (acetic acid) is a major constituent of vinegar.

When comparing inorganic chemistry with organic chemistry the following points are worthy of note:
1. most organic chemicals are covalent;
2. inorganic reactions are generally much faster than organic reactions.

Key terms

Aliphatic compounds Organic compounds which contain straight chains of carbon atoms C – C – C – C –.
Aromatic compounds Organic compounds joined end-to-end in a ring-like fashion. E.g. Benzene (C_6H_6).

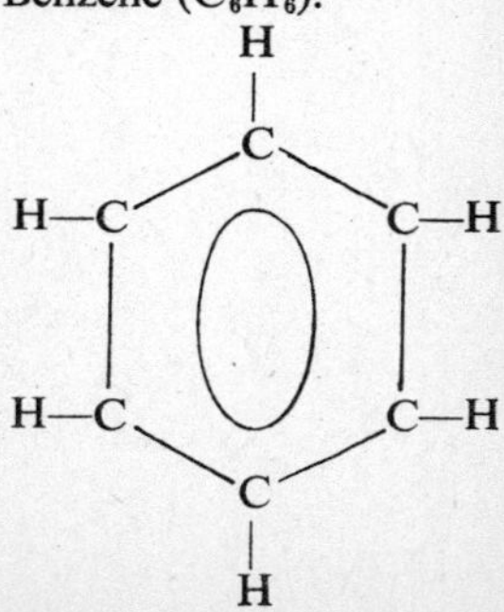

Homologous series A group of compounds which can be represented by a general formula. Adjacent members differ by CH_2.
E.g. CH_4 (methane), C_2H_6 (ethane), etc.
Alkanes A homologous series of hydrocarbons having the general formula C_nH_{2n+2}.
Alkenes A homologous series of hydrocarbons having the general formula C_nH_{2n}.
Saturated compound A compound which reacts by substitution, or contains no double ('spare') bonds.
Unsaturated compound A compound which contains 'spare' bonds and reacts by addition.
Isomerism Two or more compounds having the same molecular formula but different structural formulae are said to exhibit isomerism. Each compound is called an isomer.
Polymer A very large molecule which may have a molecular weight of many thousands.
Polymerisation A process whereby a molecule having a very high molecular weight (polymer) is made from a molecule of low molecular weight (monomer), e.g. polythene (polymer) from ethane (monomer).
Esters A series of organic compounds typified by their sweet smell, and prepared by the action of an organic acid on an alcohol.
Cracking The name given to the process of splitting up aliphatic hydrocarbon molecules into simpler substances. The process is carried out at a high temperature and a catalyst (catalytic cracking) is used.

$$\underset{\text{decane}}{C_{10}H_{22}} = \underset{\text{octane}}{C_8H_{18}} + \underset{\text{ethene}}{C_2H_4}$$

Chapter 13 Radioactivity

Discovery

Henri Becquerel in 1896 noticed that some photographic plates although previously unopened, and completely covered in light-proof paper had become fogged. Close to the plates in the drawer was a piece of a rock-like substance called pitchblende, which is an ore of the element uranium. Becquerel, after a series of experiments, eventually came to the conclusion that the uranium ore must have been emitting some kind of rays or radiation that had passed through the protective paper around the photographic plates.

Pierre and Marie Curie continued with work on similar compounds. Madame Curie called the property of emitting these rays radioactivity. They isolated two new elements, radium and polonium, both of which were radioactive.

Lord Rutherford in 1899 found that the radiation discovered by Becquerel was in fact made up of three different forms, which he named after the first three letters of the Greek alphabet, alpha (α), beta (β) and gamma (γ). These are the three types of radioactivity – α particles, β particles and γ rays.

Radioactivity can be natural or artificial. Isotopes (page 53) of most elements have atoms and nuclei that are stable. However, some of the heavier atoms have nuclei that are unstable and break down with the emission of radiation, eventually forming a new element that is stable. The 92 elements found in the Earth's crust have over 300 isotopes, and of these 50 are naturally radioactive, e.g. the three isotopes of uranium having atomic mass 234, 235 and 238, respectively, are all radioactive.

Artificial radioactive isotopes are man-made elements that are unstable, and manufactured for convenience and certain specific and various purposes.

Types of radiation

Alpha (α) particles

These are helium nuclei having an atomic mass of 4 and a charge of +2. They are not very penetrative being stopped by a thin sheet of paper.

They are written thus

$^{4}_{2}He$ or $^{4}_{2}\alpha$ or just α

Consisting of a cluster of 2 protons (each of mass 1, charge +1) and 2 neutrons (each of mass 1, charge 0), any element emitting an α particle would have its atomic mass reduced by 4 and its atomic number reduced by 2. The new element formed would occupy a place in the Periodic table two places to the left of the original radioactive element.

e.g. $^{232}_{90}Th - ^{4}_{2}He \rightarrow ^{228}_{88}Ra$

$^{232}_{90}Th - \alpha \rightarrow ^{228}_{88}Ra$

This could also be written as:

$^{232}_{90}Th \rightarrow ^{228}_{88}Ra + ^{4}_{2}He(\alpha)$

The $^{228}_{88}$ isotope of radium formed if unstable would then decay radioactively itself, and so on until a stable state is reached. See Radioactive Disintegration Series (page 186). α particles are deflected by magnetic and electrostatic fields.

Beta (β) particles

These are high-speed electrons (99 per cent of the speed of light) having zero mass, and a charge of −1 (see page 52). More penetrative than α particles, they can pass through several millimetres of metal foil before being stopped. They are written thus:

$^{0}_{-1}e$ or β

Any element emitting a β particle would have its atomic mass unchanged but its atomic number would increase by one. The new element formed would occupy a place in the Periodic table one place to the right of the original radioactive element.

e.g. $^{228}_{88}Ra - ^{0}_{-1}e \rightarrow ^{228}_{89}Ac$

$^{228}_{88}Ra \rightarrow ^{228}_{89}Ac + ^{0}_{-1}e$

Note that in this particular case a negative charge is being lost.

$88 - (-1) \rightarrow 89$

One might wonder how electrons can be lost by the nucleus of an atom. What occurs is that a neutron from the nucleus changes into a proton and an electron, thus:

$${}^{1}_{0}\text{neutron} \rightarrow {}^{1}_{1}\text{proton} + {}^{\ 0}_{-1}\text{electron}$$

i.e. $n \rightarrow p + {}^{\ 0}_{-1}e$

β particles are deflected by both magnetic and electric fields.

Gamma (γ) rays

These have no charge or mass and are similar to light and X-rays (although much more penetrative). They are the most penetrative of the three radiations, and will pass through lead sheet several centimetres thick. They are short-wavelength rays of the electromagnetic type. They are unaffected by electric and magnetic fields. Gamma rays are the highly dangerous radiation produced by the explosion of a nuclear device. Uncontrolled they can severely damage the human body.

Summary of the properties of each radiation

Radiation	Mass	Charge	Symbol of particle	Penetrating power
Alpha particle	4	+2	${}^{4}_{2}He$ or α	Stopped by a thin sheet of paper
Beta particle	0	−1	${}^{\ 0}_{-1}e$ or β	Stopped by 2 cm of aluminium
Gamma rays	0	0	γ	Stopped by lead sheet 5 cm thick

The table indicates that the least penetrative radiation (α particles) is stopped by a sheet of paper and can travel just a few centimetres unimpeded in air.

The second most penetrative radiation (β particles) is stopped by 2 cm of aluminium sheet and unimpeded can travel a few metres in air.

Most penetrative of all, γ rays are stopped by 5 cm of lead metal, and can travel between 90 and 100 m unimpeded in air.

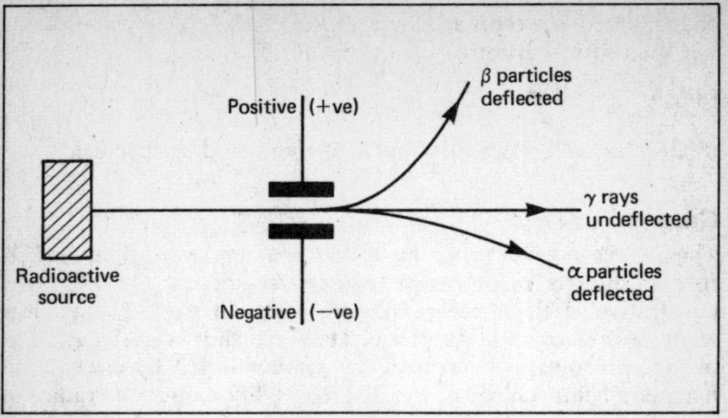

Figure 40. Effect of a magnetic field on each radiation

Half-life period of a radioactive element

Each radioactive element decays at a different rate to each other, although it is as well to point out that the rate of decay for a particular element is constant and never varies. Obviously there has to be some way of measuring the respective rates of decay.

The half-life ($t_{\frac{1}{2}}$) is the time during which half of the atoms of a particular isotope originally present disintegrate. This is independent of the original mass.

E.g. the half-life period of the isotope of lead of atomic mass 210 is 22 years. This means that 1g of this isotope in 1982 would have become 0.5g by the year 2004, i.e. half the atoms would have disintegrated by the emission of radiation. The remaining 0.5g of lead would take a further 22 years to disintegrate to 0.25g, and so on.

Half-life can also be defined as the time taken for the radioactive substance to lose half its radioactivity. Theoretically on this basis a radioactive element will never lose all its radioactivity; however, the radiation is regarded as being insignificant after seven half-lives.

As stated at the beginning of this section, the half-life period varies from isotope to isotope. Some are tiny fractions of seconds, whereas others are thousands of years.

The half-life periods of some isotopes are given in the table below.

Element	Symbol	Half-life
Lead	$^{210}_{82}Pb$	22 years
Radon	$^{220}_{86}Rn$	55 seconds
Hydrogen	$^{3}_{1}H$	12 years
Uranium	$^{239}_{92}U$	24 minutes
Carbon	$^{14}_{6}C$	5,700 years
Protoactinium	$^{234}_{91}Pa$	68 seconds
Uranium	$^{238}_{92}U$	4.5×10^9 years
Thorium	$^{234}_{90}Th$	2.41 days

Measurement of the half-life period

This is done using a Geiger–Müller counter, which is also used to detect the presence of radioactivity. The rate of radioactive decay is measured at various intervals of time. The table gives the results of an experiment in which measurements of the radioactivity of an isotope were taken.

Time (minutes)	Counts (per second)
0	800
10	560
20	400
30	280
40	200
50	140
60	100
70	70
80	50
90	35
100	25

These results were plotted on a graph (Figure 41, p. 186).

Obviously from both the table and the graph the half-life of this particular isotope is 20 minutes.
Remember that (i) radioactive elements each decay at a fixed rate known as the half-life period, and (ii) radioactivity is totally unaffected by any chemical change, or any change whatsoever in temperature and pressure, or by the effects of magnetic or electric fields.

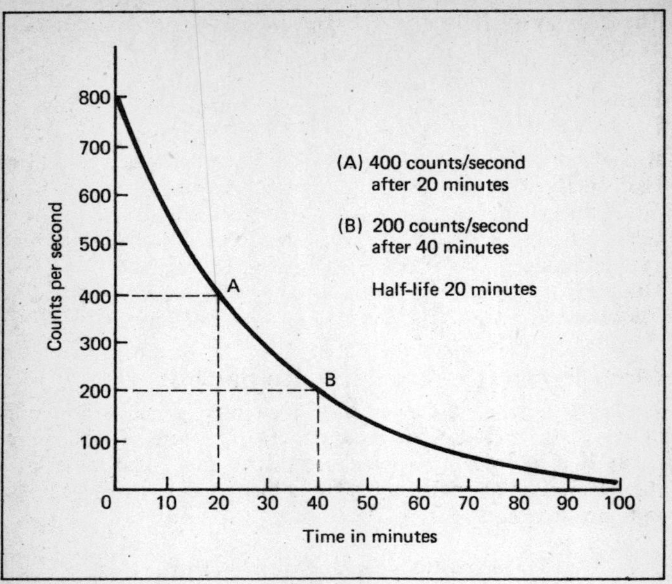

Figure 41. Graph of the rate of radioactive decay

Radioactive disintegration series

When a radioactive element decays, radiation is emitted, and a new element is formed. If the nucleus of this new element is unstable, then it itself will decay. This disintegration process will continue until a stable state is eventually reached. The following example shows the disintegration of actinium through the various stages until the stable state (lead-208 in this case) is reached.

$$^{228}_{89}\mathrm{Ac} - \beta \rightarrow {}^{228}_{90}\mathrm{Th} - \alpha \rightarrow {}^{224}_{88}\mathrm{Rn}$$
$$^{224}_{88}\mathrm{Rn} - \alpha \rightarrow {}^{220}_{86}\mathrm{Rn} - \alpha \rightarrow {}^{216}_{84}\mathrm{Po}$$
$$^{216}_{84}\mathrm{Po} - \alpha \rightarrow {}^{212}_{82}\mathrm{Pb} - \beta \rightarrow {}^{212}_{83}\mathrm{Bi}$$
$$^{212}_{83}\mathrm{Bi} - \beta \rightarrow {}^{212}_{84}\mathrm{Po} - \alpha \rightarrow {}^{208}_{82}\mathrm{Pb}$$

Uses of radioactivity

Although naturally occurring isotopes are used in certain instances, man-made isotopes or radioisotopes are often manufactured for certain and specific purposes, because of their particular radioactive and often other properties.

1. Medicine

(i) For the treatment of cancers, using γ radiation.

(ii) Radioactive iodine ($^{127}_{53}I$) is used in the treatment of thyroid gland disorders.

(iii) Radioactive gold needles are implanted in the body for the further treatment of tumours. In this particular case, it is vital to realise that gold is used because of its non-reactivity. It would be nonsensical to implant substances in the body that would actually react with the body.

(iv) Small quantities of radioactive substances can be swallowed by human beings. These substances eventually find their way into the tissues and blood stream, consequently large organs such as the lungs, liver, kidneys, etc., can be studied by the radiation emanated from them. The material swallowed must also be easily removable from the body by the natural processes of waste removal.

(v) The highly penetrative γ radiation also finds a use in medicine in order to sterilise surgical and other equipment.

2. Engine wear

The pistons, block and other parts of large industrial engines often contain small quantities of a radioactive substance, included when the parts are manufactured. Wear in the engine is indicated by the radioactivity of the oil used to lubricate the engine. Such a system saves the time of unnecessarily taking down engines to check on their condition, and helps improve the design of the engine and the efficiency of the lubricant.

3. Carbon dating

The age of wooden objects can be found by measuring their content of the isotope of carbon $^{14}_{6}C$. Very small quantities of nitrogen atoms in the atmosphere change into $^{14}_{6}C$ atoms. These radioactive carbon atoms eventually find their way into living trees by the process of photosynthesis. Obviously when a tree is felled the process ceases. By comparing the radioactivity of a modern wood with that of a wood of unknown age, the time that has elapsed since the unknown wood was cut down can be found. This method finds use in dating archaeological remains.

4. Flaw detection

Radioisotopes are used in apparatus to 'X-ray' metals and castings, etc., to detect cracks, flaws, and other imperfections invisible on the surface.

5. Detection of leaks in pipes and sewers

This is achieved by putting radioactive sodium chloride solution into the water passing down the pipe. Where the leak occurs, the soil in the vicinity becomes soaked with the radioactive salt solution, which can then be easily detected by a Geiger counter. Obviously this method is far more accurate and economical than that of mass excavation of the length of suspect pipe.

6. Thickness gauges

In the manufacture of metal sheet and foil, paper and plastics the thickness of the product depends upon the distance apart the rollers are set. Radioactive detector devices are used to ensure consistency of thickness. If the metal sheet becomes too thick or too thin, the detecting device will receive more or less radioactivity. The instrument then automatically adjusts the rollers according. Pin-hole detection in tin plate also makes use of a similar device.

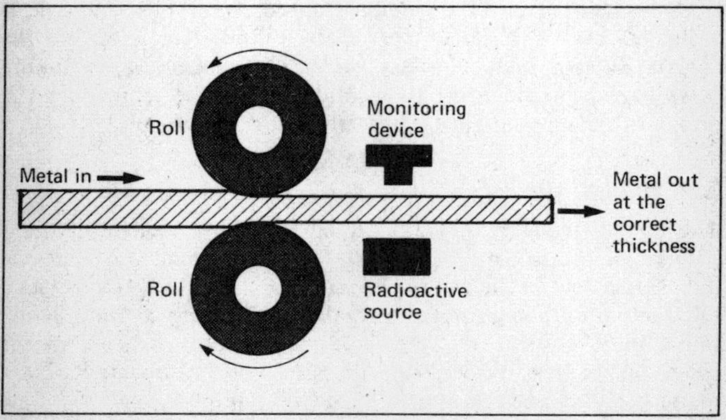

Figure 42. Illustration of a thickness gauge device

Nuclear power stations

In all power stations the fuel used heats water to produce steam. The pressure of the steam is then used to drive generators that feed electricity into the National Grid System. This principle of using steam to drive a mechanical device has been used for many years in the steam engine.

Conventional power stations use coal, oil or gas as fuels. In nuclear power stations, the heat produced by a nuclear reaction in the reactor is carried away by a gas or a liquid. This hot gas or liquid is used to boil the water in huge boilers. Steam is produced and drives the generators.

Nuclear waste

The most up-to-date method of dealing with waste from nuclear reactors is to encapsulate the waste in glass. Sealed in purpose-built containers the glass 'marbles' produced are then dropped into the deep waters of the seas. This method of dealing with nuclear waste is reported to be effective, cheap and convenient.

Key terms

Radioactivity. The property of spontaneous emission of penetrating radiation by certain unstable types of atomic nuclei. Disintegration of the nuclei is caused by the emission, which may be either α or β particles and/or γ rays.

Alpha (α) particles are fast-moving helium nuclei having a mass of 4 and a charge of +2. They are written ^4_2He or $^4_2\alpha$ or α.

Beta (β) particles are very fast-moving electrons, having zero mass and a charge of −1. They are written $_{-1}^{0}e$ or β.

Gamma (γ) rays are similar to light and X-rays with no mass or charge. They are the dangerous radiation produced during a nuclear explosion.

Half-life period is the time for half of the atoms of a particular radioactive isotope originally present to disintegrate.

Isotopes are atoms of the same element, each having the same atomic number, but different atomic masses.

Chapter 14
Qualitative Analysis

All examination boards ask questions relating to tests for various substances. A number of the tests that follow are to be found under separate headings in the book. However, it is convenient to have the tests summarised under one chapter. These can be conveniently classified into: gases; anions (negative ions); and cations (positive ions).

1. Tests for gases

Hydrogen Test the gas with a lighted taper. It burns with a pop and a blue flame. This gas forms an explosive mixture with air, and if this air–hydrogen mixture is ignited it explodes (pops in small quantities). Odourless and colourless.

Oxygen Test the gas with a glowing splint. Oxygen will relight the glowing splint. Odourless and colourless.

Carbon dioxide Test the gas by bubbling it through a solution of limewater (calcium hydroxide). A white precipitate of calcium carbonate is formed, i.e. the limewater turns milky.

$$Ca(OH)_2(aq) + CO_2(g) = \underset{\text{(causes milkiness)}}{CaCO_3(s)} + H_2O(1)$$

Odourless and colourless.

Ammonia This is the only alkali gas. It will turn moist red litmus indicator paper blue. In addition, ammonia gas will form dense white fumes of ammonium chloride with hydrogen chloride gas.

$$NH_3(g) + HCl(g) = \underset{\text{white fumes}}{NH_4Cl(g)}$$

Ammonia also has a characteristic smell ('babies nappies'), and is colourless.

Hydrogen chloride An acid gas that gives white fumes with ammonia gas (as above).

$NH_3(g) + HCl(g) = NH_4Cl(g)$

A sharp acidic smell. Colourless.

Sulphur dioxide Bubble the gas through orange potassium dichromate(VI) solution. The solution turns green. It will also decolourise an acid solution of potassium manganate(VII). It also has a characteristic smell and taste.

Chlorine Turns damp blue litmus paper red before bleaching it white. This gas will displace bromine from solutions of bromides and iodine from solutions of iodides. Starch iodide paper turns blue when tested with chlorine. Chlorine has a choking smell and is a greenish yellow colour.

Water vapour Turns blue cobalt(II) chloride paper pink and turns white anhydrous copper(II) sulphate blue. Odourless and colourless.

Hydrogen sulphide Turns clear lead nitrate solution black. It smells of rotten eggs.

Caution When testing for gases, great care should be taken not to inhale large quantities of any gas. The slightest 'whiff' of the gas will be sufficient in all cases. The results should be confirmed by using chemical tests.

2. Tests for anions (negative ions)

So called because these ions would migrate towards the positive electrode (anode) in electrolysis experiments.

Chloride

(i) In the solid state. Add concentrated sulphuric acid. Hydrogen chloride gas is evolved. This gas can be tested as described previously.

$$NaCl(s) + H_2SO_4(l) = NaHSO_4(aq) + HCl(g)$$

(ii) In solution. Add a solution of silver nitrate to a solution of the chloride acidified with dilute nitric acid. A white precipitate (which darkens in light) is formed. This precipitate dissolves readily in concentrated ammonia solution (ammonium hydroxide).

$$NaCl(aq) + AgNO_3(aq) = AgCl(s) + NaNO_3(aq)$$

Bromide

(i) In the solid state. Add concentrated sulphuric acid. Reddish brown fumes of bromine gas are evolved.

(ii) In solution. Add a solution of silver nitrate to a solution of the bromide acidified with nitric acid. An off-white precipitate of silver bromide is formed that is sparingly soluble in ammonia solution (ammonium hydroxide).

$$NaBr(aq) + AgNO_3(aq) = AgBr(s) + NaNO_3(aq)$$

Iodide

(i) In the solid state. Add concentrated sulphuric acid. Violet fumes of iodine vapour are evolved.

(ii) In solution. Add a solution of silver nitrate to a solution of the iodide acidified with dilute nitric acid. A pale yellow precipitate of silver iodide is formed that is insoluble in ammonia solution (ammonium hydroxide).

$$NaI(aq) + AgNO_3(aq) = AgI(s) + NaNO_3(aq)$$

Carbonate The same test occurs in both the solid and solution states. Add a dilute acid, and carbon dioxide is evolved.

$$Na_2CO_3(aq) + 2HCl(aq) = NaCl(aq) + H_2O(1) + CO_2(g)$$

Hydrogencarbonates It is as well to remember that there are only two solid hydrogen carbonates, those of sodium and potassium. These two can be differentiated from the corresponding carbonates by heating. When a hydrogencarbonate is heated, carbon dioxide is evolved and the carbonate remains.

$$2NaHCO_3(s) = Na_2CO_3(s) + H_2O(1) + CO_2(g)$$

All hydrogen carbonates give carbon dioxide when reacted with an acid.

Nitrate

(i) In the solid state. Mix the solid with some copper turnings. Add concentrated sulphuric acid and heat gently. Reddish brown fumes of nitrogen dioxide are evolved. The nitrate reacts with the acid to produce nitric acid.

$$NaNO_3(s) + H_2SO_4(1) = NaHSO_4(aq) + HNO_3(g)$$

The nitric acid formed then reacts with copper.

$$Cu(s) + 4HNO_3(g) = Cu(NO_3)_2(aq) + 2H_2O(1) + 2NO_2(g)$$

Reddish brown fumes of nitrogen dioxide are evolved.

(ii) In solution there are two tests available.

(a) The brown ring test. Add a freshly prepared solution of iron(II) sulphate to an equal volume of the nitrate. Carefully pour a few drops of concentrated sulphuric acid down the side of the tube. A brown ring appears at the junction of the liquids. The brown ring (see page 160) is a compound called nitroso-iron(II) sulphate $FeSO_4.NO$.

(b) Aluminium powder and sodium hydroxide are added to a solution of the nitrate and the mixture warmed. Ammonia gas is evolved.

Sulphate (SO_4^{2-}) In solution. Add a solution of barium chloride to a solution of the sulphate acidified with a few drops of hydrochloric acid solution. A white precipitate of barium sulphate is formed.

$$H_2SO_4(aq) + BaCl_2(aq) \rightarrow BaSO_4(s) + 2HCl(aq)$$

Sulphite In the solid state. Add dilute hydrochloric acid solution and heat. Sulphur dioxide gas is evolved.

$$Na_2SO_3(s) + 2HCl(aq) \rightarrow 2NaCl(aq) + SO_2(g) + H_2O(1)$$

3. Tests for cations (positive ions)

So called because these ions would migrate towards the negative electrode (cathode) in electrolysis experiments. There are three tests used to identify cations.

Flame test Some cations when mixed with a few drops of concentrated hydrochloric acid will give a distinctive colour when placed into a Bunsen burner flame by means of a platinum wire. Typical colours are given in the table.

Cation	Formula of ion	Flame colour
Sodium	Na^{+}	Bright orange/yellow (the colour of some street lights)
Potassium	K^{+}	Lilac/pale purple
Calcium	Ca^{2+}	Brick red
Copper(II)	Cu^{2+}	Dark green
Barium	Ba^{2+}	Pale green

Reaction with sodium hydroxide solution In this test the precipitates are insoluble metal hydroxides, some of which will dissolve in excess sodium hydroxide solution. Obviously because all sodium, potassium, and ammonium compounds are soluble, no precipitate will be obtained with any solutions of their compounds with sodium hydroxide. The table gives the results with some common cations.

Cation	Formula of ion	Result of adding a few drops of sodium hydroxide solution	Result of adding sodium hydroxide solution in excess
Calcium	Ca^{2+}	A very faint white precipitate is formed	No change. The precipitate is insoluble
Iron(II)	Fe^{2+}	A dirty green precipitate is formed	No change. The precipitate is insoluble
Iron(III)	Fe^{3+}	A reddish brown (rusty iron) precipitate is formed	No change. The precipitate is insoluble
Copper(II)	Cu^{2+}	A blue precipitate is formed	No change. The precipitate is insoluble
Zinc	Zn^{2+}	A white precipitate is formed	The precipitate dissolves to form a colourless solution. The precipitate is soluble

Aluminium	Al^{3+}	A white precipitate is formed	The precipitate dissolves to form a colourless solution. The precipitate is soluble
Lead	Pb^{2+}	A white precipitate is formed	The precipitate dissolves to form a colourless solution. The precipitate is soluble

Reaction with ammonium hydroxide The reactions here are very similar to those with sodium hydroxide solution, metal hydroxides being formed. There are a few notable differences. The table gives the results with some common cations. Differences in the reaction with ammonium hydroxide and sodium hydroxide are asterisked (*).

Cation	**Formula of ion**	**Result of adding a few drops of ammonium hydroxide solution**	**Result of adding excess ammonium hydroxide solution**
*Calcium	Ca^{2+}	No precipitate is formed	No change
Iron(II)	Fe^{2+}	A dirty green precipitate is formed	No change. The precipitate is insoluble
Iron(III)	Fe^{3+}	A reddish brown (rusty iron) precipitate is formed	No change. The precipitate is insoluble
Copper(II)	Cu^{2+}	A blue precipitate is formed	The precipitate dissolves to form a deep blue solution. The precipitate is soluble

Zinc	Zn^{2+}	A white precipitate is formed	The precipitate dissolves to form a colourless solution. The precipitate is soluble
*Aluminium	Al^{3+}	A white precipitate is formed	No change. The precipitate is insoluble
*Lead	Pb^{2+}	A white precipitate is formed	No change. The precipitate is insoluble

Caution Remember that chemicals, such as concentrated acids and alkalis, can be very dangerous. Take great care when using any chemicals, and always ensure that any spillages of chemicals are always cleaned up promptly.

Chapter 15
Plastics and Fibres

Plastics are the new materials of the 20th century, and we have come to rely heavily on articles made of plastics. From wall-papers to window panes, one only has to look around the home to realise the use, variety and versatility of plastic materials. These new materials are replacing the older and more traditional materials, such as wood, metal and stone, and in some instances proving to be superior to the 'older' materials. The term plastic indicates that the material can at some stage be moulded or shaped by heat or pressure or both. All plastics are man made materials, most being polymers (page 177). (Rubber is sometimes regarded as a naturally occurring plastic.) Plastics have advantages and disadvantages over the more traditional materials such as wood, metal and stone.

Advantages

1. They do not corrode in air or water, unlike some metals. They are indeed very 'atmosphere resistant' and are used extensively for outside use.
2. Plastics are coloured during the manufacturing process; consequently they do not require painting afterwards.
3. They are not subject to diseases of any kind, such as those that affect other more traditional materials, e.g. dry rot in wood, and other decaying processes.
4. Easily formed and shaped, they are overall much cheaper than the traditional materials.
5. Generally lighter than traditional materials they also possess great strength. This property of strength coupled with lightness is a very important property possessed by some plastic materials.

6. Many are non-toxic and they find great use in the manufacture of children's toys and for the safe packaging of food, e.g. 'cling film' and plastic bags for foodstuffs (crisps!!).
7. They are waterproof.
8. Many are made in fibre form and can be easily interwoven with natural materials such as wool to make excellent and durable composite materials, e.g. suits and coats are often made from a terylene–wool mixture.

Disadvantages

1. Many burn extremely easily and in so doing produce harmful gases and much smoke.
2. They are virtually indestructible and do not decompose in the atmosphere or under the ground; consequently waste plastic materials do create a serious disposal and pollution problem. This acute problem is being overcome to a certain extent by the production of biodegradable plastics. These materials are designed to decompose under certain atmospheric and other conditions.
3. Unlike some naturally occurring materials they have in some respects an uninteresting appearance.

Plastics and polymers (page 177) can be divided up into two distinct groups depending upon the way in which they behave when they are heated. These two groups are **thermoplastics** and **thermosetting plastics.**

1. Thermoplastics soften with heat and remain soft while hot. When they are cooled after heating they become relatively hard again. Further heating render this group soft again. As a result, such plastics can easily be moulded into shape when they are hot, retaining their shape and becoming harder and more rigid when they are cooled.

Examples of thermoplastics and their uses

Plastic	Uses
Polystyrene	Egg boxes, packaging jars, cartons and bottle closures. Expanded polystyrene is used for ceiling tiles, thermal insulation

Plastic	Uses
	and packaging materials for delicate equipment
Polytetrafluoroethane (PTFE, Teflon, Fluon)	Coatings, for non-stick frying pans and other kitchenware. This material is also used for bearings of all kinds
Polymethylmethracrylate (Perspex)	Glazing instead of glass and in decorative signs
Polyvinyl chloride (PVC)	Insulation for electrical cables, guttering and pipes, records, floor coverings. Fabrics for upholstery 'synthetic leather'
Polyvinylacetate	Emulsion adhesives and emulsion paints
Polypropylene	Components for cars, plastic films, ropes. Washing-up bowls
Polyethene (polythene)	Sheets, bags, pipes. When coloured it is sold under the trade name Alkathene

2. Thermosetting plastics harden with heat and remain hard. They cannot be softened by reheating.

Examples of thermosetting plastics and their uses

Plastics	Uses
Phenol – formaldehyde resins (e.g. Bakelite)	Electrical components such as plug tops and switch bases. Brake linings, adhesives

Plastics	Uses
Urea – formaldehyde	Electrical components, cavity wall insulation. Adhesives
Polyurethanes	Paints, adhesives, etc.
Epoxy resins	Adhesives (Araldite) and for protective lacquers
Polyester resins	Car-body repair kits
Melamine – formaldehyde resins	Unbreakable tablewear

Polymerisation (page 177)

Polymers can be made in two different ways: by addition polymerisation; and by condensation polymerisation.

Addition polymerisation This is referred to on page 177 and involves the chemical union of two or more molecules of the same compound (the monomer) to form much larger molecules (the polymer). Addition polmerisation always involves an unsaturated monomer, i.e. one that contains a double bond. The formation of polyethene (polythene) is an excellent example of addition polymerisation. If an ethene monomer is represented by the letter X, the polymer would be written

—X—X—X—X—X—X—X—X—X—X—X—

Polyethene chains can be up to 50 000 units long.

Examples of addition polymers

Polyethene

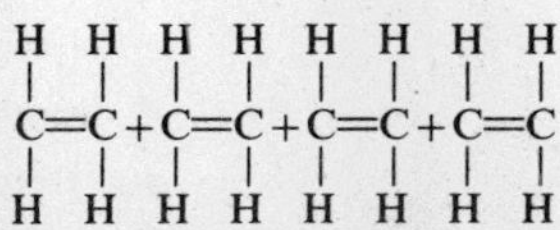

etc.

(Ethene monomers combining)

Under suitable conditions, i.e. heat, catalyst, pressure, etc., the monomers combine to form

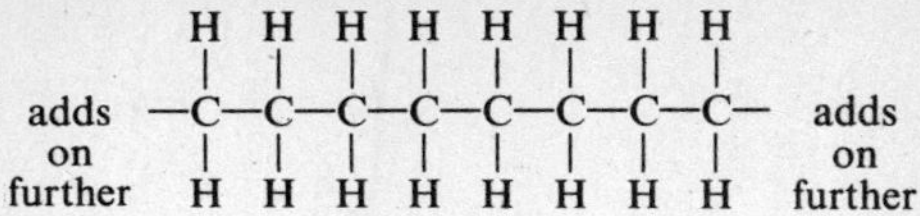

(The polymer)

At each end of the polymer there are 'spare bonds' available for further addition; consequently the length (theoretically) of the chain is limitless.

Polyvinyl chloride (PVC)

The ethene monomer consists of two carbon atoms linked by a double bond, with two hydrogens linked to each carbon. If one of these hydrogen atoms is replaced by a chlorine atom, the compound vinyl chloride is formed.

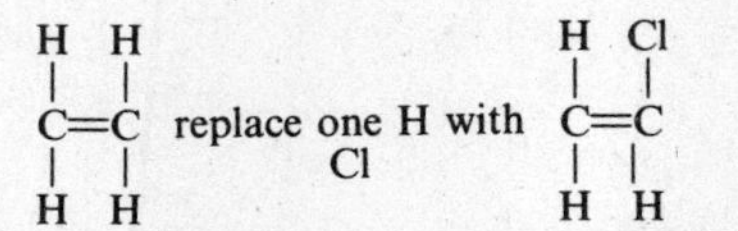

(Ethene) (Vinyl chloride)

The vinyl chloride monomer can polymerise in a similar way to ethene.

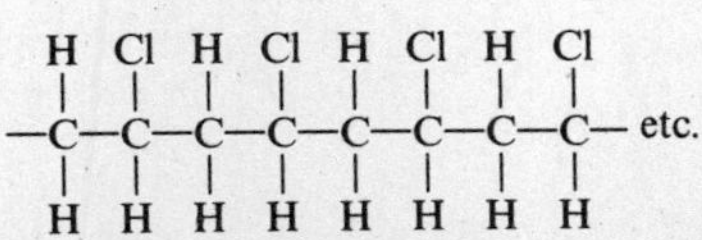

(Polyvinyl chloride) (PVC)

Polytetrafluoroethene (PTFE)

If all four hydrogen atoms in the ethene molecule are each replaced by a fluorine atom the compound tetrafluoroethene is formed.

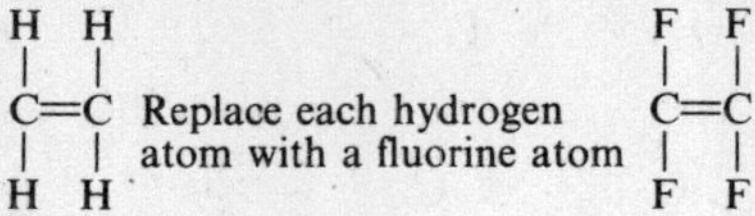

(Ethene) (Tetratluoroethene)

The tetrafluorethene monomer can polymerise in the same way as ethene.

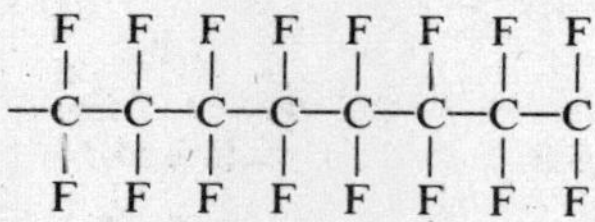

Better known as Teflon, this plastic provides the non-sticking layer on frying pans and other kitchenware, together with other uses.

Condensation polymerisation

Here two different monomers join together to form a polymer with the elimination of a small molecule, such as water (hence the name condensation polymerisation). If the monomers are represented by the letters A and B the monomer could be written:

—A—B—A—B—A—B—A—B—A—B— etc.

The polymer is formed by taking large numbers (n) of monomer A and monomer B and allowing them to combine with the elimination of water:

$$\text{i.e. } nA + nB \rightarrow (AB)_n + nH_2O$$

The formation of nylon is a good example of a condensation polymerisation reaction. Hexanedioc acid (adipic acid) ($HO_2C(CH_2)_4CO_2H$) is heated with hexan-1,6-diamine (hexamethylenediamine, HMD ($H_2N(CH_2)_6NH_2$) to form nylon 66, with the elimination of water.

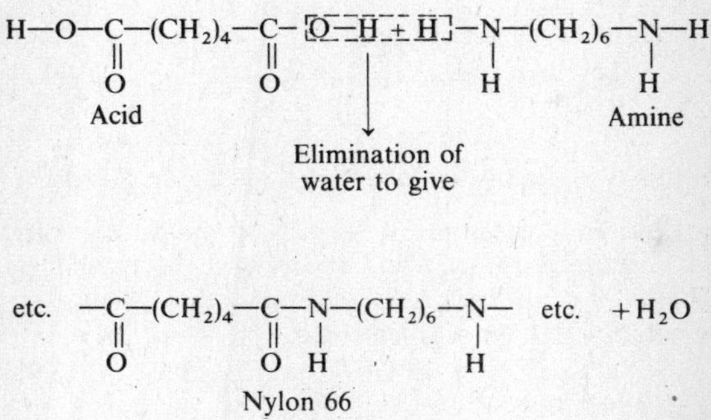

This reaction can be understood more readily by looking at the individual formulae of the acid and the amine:

Hexanedioc acid (adipic acid)

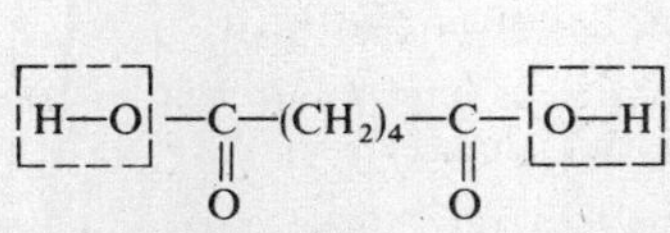

at each end of this complicated molecule is an O—H group.

Hexan-1,6-diamine (hexamethylenediamine)

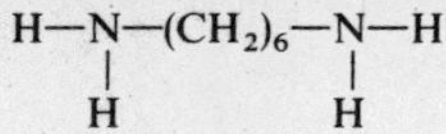

at each end of this complicated molecule is an H.
The condensation is achieved by the OH group combining with the H to complete the linking, with HOH eliminated. The fascinating part is that an OH and H remain at each end for further linking, e.g.

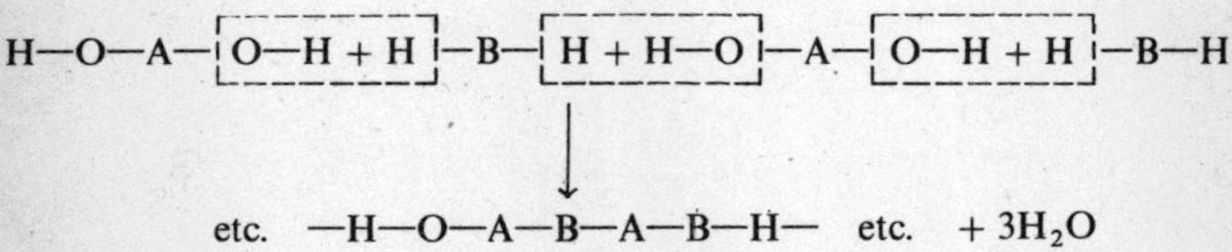

The A and B represent the acid and amine, respectively.

For examination purposes it is not required that candidates know the actual formulae, but understand the principles involved. The 'A–B–A–B' idea would be sufficient, as long as it was shown how the linkage was achieved.

Key terms

Plastic A substance that at some stage in its manufacture can be moulded or shaped by heat, pressure or both.

Thermoplastics are those that soften when heated, but become hard when cool. They can be re-softened repeatedly by heat without change of properties.

Thermosetting plastics are those that cannot be remelted again once they have solidified. That is once they have been subjected to heat, lose their plasticity.

Monomer This is a single unit (e.g. ethene) from which a polymer (polyethene) is made.

Polymer These are very long molecules that have a molecular weight of many thousands, e.g. polyethene (polythene) is formed by allowing ethene to add on to other ethene molecules under critical conditions of temperature and pressure.

Polymerisation is the process whereby a molecule having a high molecular weight (polymer) is made from a molecule of low molecular weight (monomer), e.g. polythene is a polymer, but ethene is a monomer.

Addition polymerisation Here large numbers of the same molecule combine to form a huge molecule. The polymer can be written as:

—A—A—A—A—A—A—A—A—A—A— etc.

Condensation polymerisation Here two different molecules combine to form a polymer with the elimination of a small molecule, such as water.

Index

Examination Hints

1. General advice

The basis of success in any examination lies in the **preparation** that has preceded it. 'Preparation' may usually be taken to mean **study**. Real study calls for much more active participation on the part of the student than merely sitting down and reading a book. In chemistry, it involves gradual and thorough coverage of the prescribed syllabus, trying out experiments and testing reactions in the laboratory, and a sound knowledge of chemical symbols, standard formulae and how to use equations. Regular study will prove far more productive than last-minute cramming, and will make revision at the end of the course far easier.

Well before he or she is due to take the examination, the student should know the format of the paper, e.g. whether it is in sections, whether any sections are compulsory, where there is a choice, what degree of choice is available. These facts can be ascertained by consulting the appropriate syllabus. It is dangerous to rely entirely on past papers as a guide to paper format because changes are made quite frequently. However, most of the examination papers in chemistry are comprised of two sections; Section A, which is compulsory and made up of a number of short-answer questions, and Section B, consisting of longer questions of the structured type (students are usually given a choice of questions in this section). Very few examinations nowadays are of the traditional type requiring straightforward descriptions and word-for-word regurgitation of facts. Questions are now designed to test the application of knowledge – the candidate's understanding rather than his powers of recall. In chemistry, application and evidence of understanding must be based on thorough knowledge of the facts. It is therefore of paramount importance to **understand** the facts when they are first presented, and if understanding is not achieved the student must seek out this understanding from teachers or books or both.

At the examinations, the student should first read thoroughly the instructions printed on the paper. If any particular section is compulsory it is always wise to attempt as much of this part as possible first, then turn to the other section(s). In longer-answer questions, where a choice is usually available, the student must ensure that he attempts the requisite number of questions, certainly no more and if possible no less. If a student does **more** than the required number of questions, the 'excess' questions will not be marked and obviously will not earn him any marks, e.g. if it is stipulated that four questions should be attempted and the candidate attempts five, only the **first** four questions will count towards his total mark, and **not**, as some students tend to think, the best four.

In sections where a choice is given, it is as well for the student to begin with what he thinks is his **best** question, and conclude with his 'worst'. By doing the questions in this order, he gives himself the psychological boost of starting off well, which will raise his morale for what is yet to be done.

Probably one of the major errors made by students in examinations is mis-timing the paper. As a general rule, time should be proportional to marks and *vice versa*. For example, if a paper is timed for $2\frac{1}{2}$ hours and consists of Section A (40 marks) and Section B, of which three questions are to be attempted each carrying 20 marks, the marks will be in the ratio 40:60, that is 2:3; therefore two fifths of the time (1 hour maximum) should be spent on Section A, and three fifths ($1\frac{1}{2}$ hours maximum) on Section B, i.e. $\frac{1}{2}$ hour per question. The time spent on a particular question or section must include time for checking the written answer, and if possible a quick check through the whole paper at the end of the allotted time would be advantageous.

The above recommendations can be summarised as follows:

1. Read the instructions carefully.
2. Attempt compulsory questions first.
3. Attempt 'best' questions first in sections where a choice is given.
4. Above all, watch the **timing**. (A question worth 20 marks **not attempted** means 20 marks thrown away.)
5. Check each answer on finishing it, and ultimately the whole paper.

2. Planning answers

Look before you leap. Questions should be approached in a 'scientific', or logical, manner. A rough plan of each answer should have been formed before it is written down, for valuable marks will be lost if questions are tackled in a haphazard fashion.

Consider the following question: 'Place the following metals in order of decreasing chemical activity: magnesium, iron, copper.

Justify the list by reference to
  (a) heating the metals in air;
  (b) treating the metals with dilute sulphuric acid;
  (c) adding each metal in turn to separate aqueous solutions containing the ions of one of the other metals listed.'

This is a typical question, based on the Activity series, and could yield high marks for candidates prepared to spend a few moments thinking about the answer. Common mistakes in this kind of question, caused by lack of planning, are:

1. simply writing down the facts relating to reactions and not confining the answer to the justification of the original order of activity. In this question the 'key word' is **justify**.
2. omitting chemical equations. This raises the question of the general pattern of examination answers. Here, for instance, the answer to part (a), 'heating the metals in air', should be planned thus:
   (i) what is done?
   (ii) what is seen?
   (iii) why is it seen?
   (iv) chemical equation, preferably, and where applicable the simpler ionic equation.

A brief planned answer would include:

1. reference to the extent to which the metal is heated, e.g. magnesium would burn in air after heating for a short time, whereas copper and iron, even if heated to very high temperatures, would not burn;
2. a description stating that magnesium burns with a blinding light and white magnesium oxide is formed; that iron glows and copper melts if heated to above 1083°C;
3. reference to magnesium being the most active element and having, therefore, a greater affinity for oxygen than the others. Its position at the head of the list is justified by the results and observations of the experiments.
4. the equation $Mg^{2+} + 2e = Mg$.

3. Further advice

Definitions are usually required in chemistry examinations, and provide a lucrative source of marks if known. All definitions and laws should be known and where possible examples quoted. For example:

The law of constant composition states that a given pure chemical compound always contains the same elements combined in the same proportions by weight no matter how it is made. E.g. When copper (II) carbonate is heated to completion, copper (II) oxide is formed. The composition of this oxide will be exactly the same as that made by heating copper (II) nitrate to completion.

Comparison questions are quite often included, and it is essential that the subject matter be compared point by point, and not by means of separate paragraphs on each in turn. Remember, if comparing reactions, that the conditions of the reactions are important.

Relating facts It is always helpful when learning chemistry to relate the extraction of the metal and the properties of the element and its compounds to the element's position in the Activity (Electrochemical) series. E.g. The metal sodium is extracted by electrolytic means, because the compounds of sodium are very stable so both electrical power and high temperature are needed in order to break them down. It therefore follows that the metal sodium must be reactive to be able to form such stable compounds.

Examination papers often include a question asking why aluminium cannot be extracted in a blast furnace, and also why the metal is comparatively new and expensive. The answer will be similar to the one relating to sodium. Aluminium is a more reactive metal than iron, is higher in the Activity series, and forms more stable compounds, which cannot (like iron) be reduced by carbon in a blast furnace. It can be considered 'new' because electricity is a comparatively new invention, and it is quite expensive because the extraction costs are directly proportional to the cost of electrical power.

In both questions (that on sodium and that on aluminium), there are related facts which are part and parcel of the whole answer and should be included.

Application of knowledge
It is quite possible that students may find themselves faced with a question concerning substances that they have not actually dealt with in class. However, even if they have never heard of the substance, they will always be given enough information from which to deduce the answer. For example, 'Is molybdenum

nitrate soluble or insoluble?' While students may not have heard of molybdenum, they should know that all nitrates are soluble, hence molybdenum nitrate would be soluble.

Questions which require basic knowledge, then deduction, are often centred around the Periodic table, and, in particular, the families of elements. E.g. 'The metal rubidium (Rb) is in the same group as sodium and potassium in the Periodic table, and has an atomic number of 37. From your knowledge of the chemistry of sodium and potassium deduce:

(i) how the metal should be stored;
(ii) the colour and nature of its compounds;
(iii) its action with water;
(iv) how it could be manufactured.'

Once the student has recognised the group relationship the question becomes relatively simple. As the electropositive nature of metals increases in accordance with how far down the group they appear, rubidium will be more reactive than potassium. This deduction is further confirmed by the fact that potassium is more reactive than sodium, hence logically rubidium should be more reactive than potassium.

Establishment of the facts listed above makes the following answers possible.

(i) Rubidium should be stored under oil in an airtight container, preferably made of aluminium (a glass container would break if dropped). Both sodium and potassium are stored in this way owing to their reactivity, so similar treatment could be recommended for rubidium.

(ii) Most sodium and potassium compounds are white, crystalline, ionic and therefore stable compounds. Rubidium compounds would be similar, except that they would be more stable (because rubidium is more reactive than the other two metals).

(iii) Potassium burns on water, and the reaction is violent. Rubidium would probably react with explosive violence, displacing hydrogen from the water. Rubidium hydroxide, an alkali, would be formed.

$$2K + 2H_2O = 2KOH + H_2\uparrow$$
$$2Rb + 2H_2O = 2RbOH + H_2\uparrow$$

(iv) Rubidium would probably be manufactured by electrolytic means owing to the stability of its compounds. Compare with the manufacture of sodium and potassium.

4. Calculations

The majority of students have trouble with chemical calculations. Even those who are able to tackle them often spoil their work through arithmetical errors. Unfortunately an arithmetical slip made early on in a question can lead to difficulties later, so it is worth paying special attention to the arithmetic involved. Logarithm tables or slide rules are allowed in the examination room, but usually calculations are so designed that the numbers are 'easy' and tables therefore unnecessary. It is unwise to write calculations purely in terms of figures. Clear written statements should be made indicating to what the figures refer.

One of the most useful weapons for tackling calculations is a working knowledge of simple proportion. This principle can be applied in many circumstances. E.g. 'What fraction of a mole of sodium carbonate is contained in 250 cm^3 of 2.0 mol/l (2M) sodium carbonate (Na_2CO_3)?'

Answer In 1000 cm^3 of 1.0 mol/l sodium carbonate there would be 1 mole of sodium carbonate.

In 1000 cm^3 of 2.0 mol/l sodium carbonate there would be 2 moles of sodium carbonate.

In 250 cm^3 of 2.0 mol/l sodium carbonate there would be $\frac{1}{4} \times 2$ mol/e sodium carbonate, i.e. $\frac{1}{2}$ mole of sodium carbonate.

Often it is necessary to convert grams into moles and *vice versa*. E.g. 'How many moles of carbon would contain the same number of atoms as 46g of sodium?

($A_r(Na) = 23$; $A_r(C) = 12$)'

In preparing the answer, it is important to realise that 1 mole of carbon will contain the same number of atoms as 1 mole of sodium, etc. Here, the question is 'how many moles?' and the answer – a number of moles – should appear on the right-hand side of the expression.

I.e. **Answer** 23g of sodium represent 1 mole
1g of sodium represents 1/23 mole
46g of sodium represent $1/23 \times 46$
i.e. 2 moles of sodium.

Therefore 2 moles of sodium will contain the same number of atoms as 2 moles of carbon. Answer = 2.

The above example illustrates simple proportion and the principle

of converting grams into moles, for which the general rule is to divide the number of grams by the atomic weight.
E.g. In 46g of sodium (atomic weight $A_r(Na) = 23$) there will be $\frac{46}{23}$ moles.

To convert moles into grams, multiply by the atomic weight.
E.g. 'How many grams of carbon are there in 2 moles of carbon ($A_r(C) = 12$)?'

1 mole contains 12g
2 moles contain $12 \times 2 = 24$g.

Note Care should be taken, especially in those involving the mole concept, to express calculations correctly, i.e. so that the answer appears to the right of the equation.

Calculations from equations

$NaOH + H_2SO_4 = NaHSO_4 + H_2O$

From the above equation certain facts can be deduced, i.e. 1 mole of NaOH + 1 mole H_2SO_4 **produce** 1 mole $NaHSO_4$ + 1 mole of water. If the atomic weights are known, these facts can be converted into numbers.

What is important is the **ratio** of one thing to another in the equation.
E.g. $Na_2CO_3 + 2HCl = 2NaCl + H_2O + CO_2\uparrow$
From this equation:

1 mole Na_2CO_3 + 2 moles HCl produce 2 moles NaCl + 1 mole H_2O + 1 mole CO_2

Consider the following question:
'What volume of 0.15 mol/l hydrochloric acid is needed to completely dissolve 1.06g of anyhydrous sodium carbonate ($A_r(Na) = 23$; $A_r(C) = 12$; $A_r(O) = 16$)?'
To formulate an answer, first of all write down the equation:

$Na_2CO_3 + 2HCl = 2NaCl + H_2O + CO_2$

and make sure it is balanced. Remember that the question is concerned with sodium carbonate and hydrochloric acid.
From the equation, it is apparent that:

1 mole Na_2CO_3 would require 2 moles of HCl for complete reaction.

1 mole Na_2CO_3 contains (weighs) 106g, while 2 moles of HCl contain 2000cm^3 of 1.0 mol/l HCl.

The following statement can now be made:

$$106g\ Na_2CO_3 = 2000cm^3 \text{ of } 1.0 \text{ mol/l HCl}$$

$$1g\ NaCO_3 = \frac{2000}{106} cm^3 \text{ of } 1.0 \text{ mol/l HCl}$$

$$1.06g\ Na_2CO_3 = \frac{2000}{106} \times 1.06\ cm^3 \text{ of } 1.0 \text{ mol/l HCl}$$

$$= \frac{2000}{100} = 20\ cm^3 \text{ of } 1.0 \text{ mol/l HCl}$$

The question asks for the volume of .015 mol/l HCl. It is therefore necessary to convert the 20 cm^3 of 1.0 mol/l HCl into .015 mol/l HCl. Hence,

$$20\ cm^3 \text{ of } 1.0 \text{ mol/l HCl} = x\ cm^3 \text{ of } 0.15 \text{ mol/l HCl}$$

$$x = \frac{20 \times 1}{0.15}$$

$$= 133.3\ cm^3 \text{ of } 0.15 \text{ mol/l HCl}$$

Miscellaneous calculations

Percentage composition A frequent error in this type of calculation occurs when the student calculates the percentage (by weight) of water of crystallisation in a **hydrated** salt. All too often the student uses the molecular weight of the anhydrous salt.

E.g. 'Find the percentage of water of crystallisation in washing soda, $Na_2CO_310H_2O$ (A_r(Na) = 23; A_r(C) = 12; A_r (O) = 16; A_r(H) = 1).'

The molecular weight of the hydrated sodium carbonate is:

$$\underset{Na_2}{(23 \times 2)} + \underset{C}{(12)} + \underset{O_3}{(16 \times 3)} + \underset{10H_2O}{(10 \times 18)} = 286$$

286g of hydrated sodium carbonate contain 180g of water.

$$\text{Percentage of water} = \frac{180}{286} \times 100 = 62.9\%.$$

Data questions

These questions entail making deductions from supplied data relating to certain substances.

E.g. Below is given information relating to substances P, Q, R, S and T (these are not their chemical symbols).

Substance	Melting point	Electrical conductivity (i) Solid	(ii) Fused (melted)	Effect of heating in air/oxygen
P	60°C	good	good	Burns violently. A white solid remains
Q	113°C	none	none	Burns to form a gas
R	768°C	none	good	No reaction
S	−134°C	none	none	Burns. Two gases are formed, one of which condenses to a liquid
T	1083°C	good	good	Forms a black solid which is soluble in acids.

The questions asked might be:

(i) Which of the above substances could be metals?
(ii) Which of the above substances could be compounds?
(iii) Which substance is **not** a covalent compound?
(iv) Which substance could be non-metallic?
(v) Which substance could be a hydrocarbon?

Note This sort of question is valuable in that it tests the application of knowledge rather than straight recall.

Here is a general guide to answering the above questions.

(i) Metals **generally** have high melting points. However, of much greater importance is that they are good electrical conductors both in the solid and melted states. Not all metals burn: however, the majority are oxidised when heated in air/oxygen.
The student must, in the light of this knowledge of the properties of metals, study the data provided. Substances P and T satisfy the requirements, despite the fact that P has a low melting point and could be a metal such as sodium or potassium, both of which have low melting points.

(ii) Compounds do not conduct in the solid state. When heated they would if affected yield at least two products. Electrovalent compounds have high melting points and would conduct in the liquid (fused) state, while covalent compounds would not conduct in either state. According to the given data, substances S and R satisfy the requirements.

(iii) If a compound is not covalent, it is logical to assume that it must be electrovalent. Electrovalent compounds have high melting points and would conduct in the liquid (fused) state. When heated they would either be unaffected or break down with some difficulty. Substance R is the obvious choice.

(iv) Non-metallic elements do not conduct in the solid or liquid states. The choice is therefore between substance Q and substance S. The answer is Q, because when heated in air it produces a single gaseous product, whereas S produces **two** products of combustion and would therefore be made up of at least two elements.

(v) Substance S is the hydrocarbon because it does not conduct in the solid or liquid states (its melting point indicates that it is obviously a gas), and it forms two gaseous products on heating in air/oxygen, one of which condenses to a liquid (probably water, formed from the hydrogen in the compound).

A data question could also take the form of the following question, in which the volume of a gas is evolved in a chemical equation.

E.g. 'Calculate the volume of carbon dioxide evolved when 4.2g of sodium hydrogencarbonate is heated to constant weight
(molar volume of a gas at s.t.p. $= 22.4 \, l \, mol^{-1}$)
($A_r(Na) = 23; A_r(C) = 12; A_r(O) = 16; A_r(H) = 1$)'

The molar volume of a gas at s.t.p. is $22.4 \, l \, mol^{-1}$ (l/mol). This means that 1 mole of any gas at s.t.p. occupies a volume equal to 22.4 litres.

E.g. 48g of CO_2 (1 mole) would occupy 22.4 l at s.t.p.

Answer Write down the chemical equation first:

$$2NaHCO_3 = Na_2CO_3 + CO_2\uparrow + H_2O$$

The equation tells us that:

2 moles of $NaHCO_3$ produce 1 mole of CO_2 upon heating, i.e. 2×84g $NaHCO_3$ produce 22.4 l CO_2 at s.t.p.

$$1g \; NaHCO_3 \text{ would produce } \frac{22.4}{2 \times 84}$$

$$4.2\text{g } NaHCO_3 \text{ would produce } \frac{22.4}{2 \times 84} \times 4.2 \text{ l of } CO_2 \text{ at s.t.p.}$$
$$= 0.56 \text{ litres.}$$

Empirical formulae

The empirical formula of a compound shows the ratio of the numbers of atoms of each element present in one molecule of the compound. The **molecular formula** expresses the exact number of atoms of each element present in a compound. E.g. If the empirical formula for a compound were calculated to be $Na_4C_2O_6$, in order to find the molecular formula it would be necessary to know the molecular weight, i.e. 106, therefore the molecular formula would be Na_2CO_3. The calculation of the empirical formula of a compound is fairly easy and entails:

1. dividing the percentage composition by the atomic weight of each element;
2. dividing by the highest common factor, i.e. the lowest weight, so that the relative numbers of atoms of each element can be expressed in whole numbers.

E.g. A compound contains 40% carbon, 6.7% hydrogen and 53.3% oxygen. Calculate the empirical formula of the compound ($A_r(C) = 12$, $A_r(O) = 16$, $A_r(H) = 1$).

The first step is to divide each composition by its relevant atomic weight.

carbon	hydrogen	oxygen
$40 \div 12$	$6.7 \div 1$	$53.3 \div 16$
3.33	6.7	3.33

The ratio of carbon: hydrogen: oxygen in whole numbers is now calculated by dividing by the smallest, i.e. 3.33, thus:

carbon	:	hydrogen	:	oxygen
$\frac{3.33}{3.33}$	:	$\frac{6.7}{3.33}$	:	$\frac{3.33}{3.33}$
1	:	2	:	1

∴ the empirical formula is CH_2O.

As stated above, it is essential to know the molecular weight, in order to find the molecular formula.

Remember the empirical formula expresses the **ratio** of the various atoms in the molecule. It does not tell us the exact number of each present, except where the empirical formula and molecular formula are the same.

5. Attention to detail

Chemistry is a practical subject, therefore inevitably questions on practical work are included in examination papers. When writing up a 'practical' or answering a question relating to an experiment it is important to pay the utmost attention to **practical details**. It is a relatively easy matter for an examiner to find out from students' answers whether practical work has been carried out properly – or done at all.

One popular practical question relates to the preparation of salts.

E.g. Describe, giving practical details, how a pure dry sample of lead sulphate could be prepared in the laboratory.
The key words in this question are:
1. practical details;
2. pure dry sample.

Students should first decide, for any question relating to salt preparation, whether the salt is **soluble** or **insoluble**. In this particular case, it must be realised that lead sulphate is insoluble (all lead salts are insoluble except lead nitrate) therefore it is prepared by the method of **double decomposition**. This involves the interaction of two solutions of soluble salts, precipitating the insoluble lead sulphate. Suitable solutions are lead nitrate and dilute sulphuric acid.

Remember to write down:
1. what you do;
2. what you see;
3. why you see it;
4. the chemical equation.

Answer A solution of dilute sulphuric acid is carefully added to a solution of lead nitrate contained in a test tube.

A white precipitate of insoluble lead sulphate is formed. The precipitate is allowed to settle, and the supernatant liquid (the liquid on the top of the precipitate) tested with a few more drops of dilute sulphuric acid to ensure that all the lead has been precipitated.

The insoluble lead sulphate is filtered off (or centrifuged) and the precipitate washed with distilled water to free it from nitric acid (which is the other product of the reaction) and any excess sulphuric acid. The precipitate is then **dried** in an oven.

$$Pb(NO_3)_2 + H_2SO_4 = PbSO_4 \downarrow + 2HNO_3$$

In this reaction double decomposition has occurred with the formation of insoluble lead sulphate.

The above answer provides a guide to the sort of detail required. Far too many students forget to mention that it is necessary (i) to test for incomplete precipitation; (ii) to wash the precipitate free from any contaminant; (iii) to dry the precipitate.

Another common examination question is based on the well-known experiment:

alkali + acid = salt + water

Remember that for this a **burette** and **pipette** are required. Remember also that the acid **neutralises** the base, and that to do this the acid is **titrated** against the base. An **indicator** is required. All these emphasised words would be used in a full answer to a question on this subject.

Questions on salt preparations are frequently set, and careful revision of all the methods of preparation is essential.

Conclusion

Students **must** prepare thoroughly for the examination: there is no easy way or short cut to success. Consultation of past question papers will give a very useful guide to the type and standard of questions which have formerly been set. Some students try to ascertain whether a pattern has been established in the questions set over recent years; however, this is a dangerous practice, for the format of examination papers is under constant scrutiny and may very well be subject to revision for the forthcoming session. Certain major topics **do** occur on papers year after year, nonetheless, and time spent revising these will not be wasted.

The greatest benefit from past papers can be derived from attempting the questions, then assessing answers by reference to notes and textbooks. Alternatively, one could select a question and prepare a skeleton answer by reference to notes and textbooks, learn the relevant points, then attempt the question under examination conditions. Nearer the examination it is invaluable to try some complete 'dummy runs', i.e. working complete papers under examination conditions.

At the examination, the student should endeavour to think positively: rather than worrying about what he/she does not know, he should apply his thoughts to the questions he **has** covered. The examiner is interested in what **is** known. Rarely can a student answer every question.